Heat Transfer 3

Mathematical and Mechanical Engineering Set

coordinated by
Abdelkhalak El Hami

Volume 11

Heat Transfer 3

Convection, Fundamentals and Monophasic Flows

Michel Ledoux
Abdelkhalak El Hami

iSTE

WILEY

First published 2022 in Great Britain and the United States by ISTE Ltd and John Wiley & Sons, Inc.

ISTE Ltd
27-37 St George's Road
London SW19 4EU
UK

www.iste.co.uk

John Wiley & Sons, Inc.
111 River Street
Hoboken, NJ 07030
USA

www.wiley.com

Library of Congress Control Number: 2022938988

British Library Cataloguing-in-Publication Data
A CIP record for this book is available from the British Library
ISBN 978-1-78630-690-6

Contents

Preface

Thermal science is to thermodynamics as decree is to law. It answers the following question – which all good leaders must (or should) ask themselves whenever they have an "idea": "How would this work in practice?". In a way, thermal science "implements" thermodynamics, of which it is a branch. A thermodynamics specialist is a kind of energy economist. Applying the first principle, they create a "grocery store". With the second principle, they talk about the quality of their products. I add or remove heat from a source or work from a system. And the temperature, among other things, defines the quality of the energy for me.

But *by what means do I take or do I give*? Even calculations of elementary reversible transformations do not tell us by what process heat passes from a source to a system.

Thermal science specifies how, but "evacuates" work. If in a given problem related to, for example, a convector where an electrical energy (therefore in the "work" category) appears, it is immediately dissipated into heat by the Joule effect.

Three heat transfer modes can be identified: conduction and radiation – which can be seen separately, although they are often paired up – and convection, which is by nature an interaction of fluid mechanics and conduction.

Dividing the study of thermal science into three is the result of logic. Presenting this work in four volumes is somewhat arbitrary; in our opinion, however, this split was necessary in order to keep the volumes in the collection a reasonable size.

This book is Volume 3 of a collection of problems on heat transfer, devoted to the fundamentals of convective transfers. Various angles of approach are addressed: empirical relations; the analytical approach of parietal phenomena (boundary layers), including the approach of integral methods and the numerical approach. The problem of exchangers is presented without claiming to be an exhaustive treatise. Other practical aspects (two-phase, phase change materials, etc.) are dealt with in Volume 4.

This work is intended to reach a wide audience, from technicians to engineers, to researchers in many disciplines, whether physicists or not, who have a one-off transfer problem to resolve in a laboratory context. With this in mind, the theoretical developments in the text itself are as direct as possible. Specialist readers, or those who are simply curious about further theoretical developments (general equations, particular problems, mathematical tools, etc.) may refer to the Appendices.

Volume 3, primarily devoted to "traditional" approaches (analytical treatment) to convection, will be of interest primarily to readers who are looking for "simple" prediction methods.

This work has four chapters.

Chapter 1 introduces some definitions and sets up the scope of this book.

Chapter 2 is devoted to the empirical approach to convection. It contains, in particular, an application limited to the most usual relationships.

Chapter 3 is divided into two parts. The first develops the boundary layer theory, the physical basis of parietal transfers. The second describes the integral methods.

Chapter 4 establishes the two main methods for the analytical design of exchangers.

The detailed calculation of the establishment of the fundamental equations is a rather important point. In order to lighten the text and to facilitate its reading by those who are not first order theorists, some points of the theory have been referred to in **Appendix 4** and **Appendix 5**. **Appendices 1**, **2** and **3** are a collection of physical data and **Appendix 6** recalls the values of the function $erf\ (x)$ and its associates, which are sometimes useful for some calculations.

July 2022

Introduction

I.1. Preamble

Thermal energy was probably first perceived (if not identified) by humanity, through the Sun. The themes of night and day are found at the center of most ancient myths. Humanity's greatest fear was probably that the Sun would not return again in the morning. Fire became controlled in approximately 400,000 BP. Thermal transfer was therefore a companion of *Homo ergaster*, long before *Homo sapiens sapiens*.

However, it took a few hundred thousand years before so-called "modern" science was born. Newtonian mechanics dates from three centuries ago. Paradoxically, another century and a half passed by before energy was corrections perceived by scientists, in terms of the new field of thermodynamics. Furthermore, a systematic study of heat transfer mechanisms was carried out at the end of the 19th century, and even later for the study of limit layers, the basis of convection.

Heating, lighting and operating the steam engines of the 19th century were all very prosaic concerns. Yet this is where revolutions in the history of physics began: the explosion of statistical thermodynamics driven by Boltzmann's genius, and quantum mechanics erupted with Planck, again with Boltzmann's involvement.

Advances in radiation science, particularly in sensor technology, has enabled us to push back our "vision" of the universe by a considerable number of light years. To these advances we owe, in particular, the renewed interest in general relativity that quantum mechanics had slightly eclipsed, through demonstration of *black holes*, the physics of which may still hold further surprises for us.

Closer to home, fundamental thermal science, where it is conduction, convection or radiation, contributes to the improvement of our daily lives. This is particularly true in the field of housing where it contributes, under pressure from environmental questions, to the evolution of new concepts such as the active house.

The physics that we describe in this way, and to which we will perhaps introduce some readers, is therefore related both to the pinnacles of knowledge and the banality of our daily lives. Modestly, we will place our ambition in this latter area.

There are numerous heat transfer textbooks in different formats: “handbooks” attempting to be exhaustive are an irreplaceable collection of correlations. High-level courses, at universities or engineering schools, are also quite exhaustive, but they remain demanding for the listener or the reader. Specialist, more empirical thematic manuals are still focused on specialists in spite of all this.

So why do we need another book?

The authors have taught at university level and in prestigious French engineering schools, and have been involved in the training of engineers on block-release courses. This last method of teaching, which has been gaining popularity in recent years, particularly in Europe, incorporates a distinctive feature from an educational point of view. Its practice has, in part, inspired this book.

The aim is to help learners who have not had high-level mathematical training in their first years following the French Baccalaureate (therefore accessible to apprentices), and pupils with more traditional profiles. At the same time, we would like to show this broad audience the very new possibilities in the field of digital processing of complex problems.

When a miner wants to detach a block of coal or precious mineral from a wall, they pick up a pneumatic drill. If we want to construct a tunnel, we must use dynamite. The same is true for physicists.

Whether they are researchers, engineers or simply teachers, scientists have two tools in their hands: a *calculator* and a *computer* (with very variable power). Since both authors are teacher researchers, they know they owe everything to the invention of the computer. From the point of view of teaching, however, each one of the two authors has remained specialist, one holding out for the calculator and “back-of-the-napkin” calculations, and the other one using digital calculations.

The revolution that digital tools has generated in the world of “science” and “technical” fields, aside from the context of our daily lives, no longer needs to be proved. We are a “has been” nowadays if we do not talk about Industry 4.0. The “digital divide” is bigger than the social divide, unless it is part of it…

Indeed, the memory of this revolution is now fading. Have students today ever had a "slide rule" in their hands? Do they even know what it is? Yet, all the physicists behind the laws of thermal science had only this tool in hand, giving three significant figures (four with good visibility and tenacity), leaving the user to find a power of ten of the result. It goes without saying that a simple calculation of a reversible adiabatic expansion became an ordeal, which played a part in degrading the already negative image of thermodynamics held by the average student.

This reminder will seem useless to some; slide rules are at best sleeping in drawers. But there is a moral to this story: no matter what type of keyboard we type on, a calculator or a computer, our head must have control over our fingers. This book has been written on the basis of this moral.

A good physicist must have a perfect understanding of the idea of an "order of magnitude". For this, the tool is a calculator. We always do a rough sizing of a project before moving on to detailed modeling and numerical calculations.

The two authors belong to the world of engineering sciences, meaning most of their PhD students have entered the private sector. One of them, having moved into the aerospace sector, came back to see us very surprised by the recurrence of "back-of-the-napkin" calculations in his day-to-day work.

Fundamental or "basic physics" concepts are taken from a type of manual that is resolutely different from those dedicated to the numerical approach. In this case, the authors allow themselves to believe that it is no bad thing to collect them all together in a single book, for once. This is a significant difference that will surprise some and, without doubt, be criticized by others. Nevertheless, when reading this book, an "average" student will be initiated to a field that teaching models generally promised "for later on" (or never if they never go beyond a certain level of education). It is also true that fully immersed in equations and complex calculations, specialist readers will be able to "be refreshed" when faced with the short exercises, which can sometimes surprise and encourage them – why not – to go back to their roots (assuming they had indeed been there).

Another significant difference is that this book is directed at a large scientific audience, which covers possibly the entire field: researchers, PhD students or those who have obtained Confirmation and are just starting out in the field, technicians, students or professionals, engineers. This last type of scientist is perhaps the main target of this book.

Dividing the study of thermal science into three volumes is the result of logic. Presenting this work in four volumes is somewhat arbitrary; in our opinion, however, this split was necessary in order to keep the volumes in the collection a reasonable size.

The first volume, entitled **Heat Transfer 1**, is dedicated to "classic" approaches (analytical treatment) to conduction, which will be of greater interest to readers who are looking for "simple" prediction methods.

The second volume, entitled **Heat Transfer 2**, is dedicated to "classic" approaches (analytical treatment) of radiation, and assembles digital approaches of these various transfer modes. It is aimed at engineers or researchers who want to resolve more complex problems.

The third and fourth volumes, entitled **Heat Transfer 3** and **Heat Transfer 4** are focused on convection transfers. **Heat Transfer 3** deals with the fundamentals, integrating various modes of approach, both empirical and theoretical (boundary layer), and gives an introduction to the theory of exchangers. As we have already pointed out, all of these transport operations are rarely pure and lead to problems that involve three inter-connecting transfer modes, conduction, convection and radiation.

Heat Transfer 4 aims to broaden the reader's horizon to more complex transfer modes, such as two-phase transfers. It looks at mass transfer, often in analogy with heat transfer, and explores less-known fields, such as phase change materials. It also introduces the electro-thermo-mechanical modeling of systems.

So, what is this book for?

Above all, it **contains problems to be worked on**, of which most are accessible to all, from the level of an apprentice technician upwards, either one or two years after the Baccalaureate. This book was written in France, where scientific teaching is structured around universities, engineering Grandes Écoles, engineering training through apprenticeships and two types of technician training sections at high schools or universities. In countries with simpler models, readers should also find it useful.

It seems necessary to surround these problems with strong reminders of past learning, so that the reader does not need to permanently refer back to their manuals. We see two advantages in this: a presentation of the scientific material focusing on the problems, and a second chance for readers to integrate notions that perhaps had not been well understood in the initial teaching.

Lastly, upon rereading, the authors also recommend this book as an introduction to the taught disciplines.

I.2. Interlude

Before our readers immerse themselves in a text that, despite our best efforts, remains intellectually demanding, we propose a short text that is a little lighter.

This does not mean that it is not significant in terms of understanding the physics behind all the calculations proposed in this book.

Let us imagine, in a "B movie" context, a somber hostel in the gray fog of a port in the middle of nowhere. Sailors from a faraway marina come and drink away their troubles. And as always, the drink helping them along, they turn to fighting.

Let us entrust Ludwig Boltzmann to the direct the film. Our B movie heroes are getting agitated, delivering blows to one another. Each one of them has moderate kinetic energy, distributed heterogeneously among them in the room. For some reason, they get involved in a general brawl. Their average kinetic energy becomes much greater.

In everyday language, we would say that ***things are hotting up.***

This would bring us right into line with a fundamental concept of Boltzmann, who was the first to hypothesize that **heat is made up of molecular agitation**. *The temperature in a gas is proportional to the average quadratic energy of the molecules that make it up.*

$$E_C = \frac{1}{2} kT$$

Using this model, we will return to the physical basis for all transport phenomena.

On the way, we rarely escape from the explosion of a door or a window, giving in under the repeated beatings of the brawlers.

We have just modeled the pressure, due to the transfer of the quantity of movement on the surface, by the impact of molecules.

Let us now imagine that the altercation is initially located in the corner of the room: a smaller group starts fighting between themselves.

From kicks to punches, after multiple impacts within the group and its immediate neighbors, the agitation will spread: *we have just seen the mechanism of heat propagation by transfer of impacts.*

Let us place an imaginary separation (geometrically but immaterially defined) at the center of the room. Let us count the sailors that cross through it within a unit of time.

This wall is now crossed by kinetic energy: we have defined a flow of heat.

Let us put a metal ring with a surface area of $S = 1\,m^2$ in the room. On both sides of this ring, the blows exchanged constitute a transfer of kinetic energy – we have just defined **the heat flow density**.

And we have just understood the nature of the propagation of thermal flows by impacts.

Let us suppose that the great majority of the brawlers come from a ship with a white uniform. Let us suppose that another boat in the port has uniforms that are red. The red ones are initially all united. We will then quickly see that the red mariners, as they receive and return blows, spread out across the room. *We have just shown the mechanism of* ***diffusion of matter****, of a component within a mixture.*

We will have a better qualitative understanding that the fundamental law of conduction (Fourier Law) is formally identical to the law for the diffusion of mass (Fick Law).

Let us put our agitated sailors in the compartments of a flatcar train, where they continue to fight. And let us start the train moving. The kinetic energy that they contain is transported from one point to another.

We have just invented ***thermal convection****.*

We can go further.

Let us imagine a series of flatcar trains on a set of parallel tracks. The train furthest to the side is fixed to a platform. All of these trains are full of sailors. Let us suppose that our train follows the outside, parallel rail tracks. No brakes will prevent these trains from moving. Only the last train, at the platform, is stuck.

For a reason we do not need to analyze (cinema allows all kinds of fantasy), "clusters" of fighting sailors jump from one wagon to the next. These "clusters" contain a component of speed that is parallel to the train, which will communicate information about the quantity of movement to the adjacent train. These trains will then start to move, more quickly the closer they are to the outside train. And the same occurs up to the train at the platform. This train will not move, but a force will be applied to its brakes.

We have just discovered the mechanism of dynamic viscosity. At the same time, the parallel trains in relative movement give us a picture of the *notion of boundary layers.*

At the same time, these agitated clusters carry their disordered kinetic energy, "thermal" agitation. We have just seen the mechanism of the *thermal boundary layer.*

Finally, let us include a few red mariners in the crowd of white. They will be carried with the clusters, and we have just invented *the limit layer of diffusion of a species.*

We are in a fantasy, and let us benefit from it as far as we can. To finish, let us suppose that this is carnival day; each sailor has a belt equipped with bells.

All the individuals have a different speed, and the impacts are random, all the bells start to jingle, each with a different frequency. The distribution of frequencies will depend on the statistical distribution of speeds (Boltzmann statistics), and the intensity of noise produced will depend on the total agitation energy of the sailors.

We have just understood the ***basic mechanism of radiation****. We have just realized why the theory of radiation needed to use the concepts of statistics from the work of Boltzmann – a brilliant pupil of Planck – to produce the emissions spectrum of a black body, for example.*

NOTE.– *The model is certainly simplistic. The emission comes from quantum transitions in the gas atoms.*

Here, we have already deviated from the pure substance of the book, but we could go even further.

Let us suppose that our agitated sailors are in a room with one mobile wall (a nightmare scenario frequently seen on the silver screen).

The incessant impacts of the fights on this wall create a force that pushes it. This force, reduced to a surface unit, explains the notion of pressure.

By pushing against this wall, our crowd applies work that is greater than the resistance.

Here we see an equivalence spring up between work and heat that, at a fundamental level, are simply two mechanical energies: one ordered and the other disordered. **The first principle of thermodynamics is illustrated by this.**

We can see that the incidence of an average blow on the wall is rarely normal.

Therefore, an average fighter will have a trajectory that will be reflected off the wall. And only the normal component of its speed will be able to push (or transfer work to) the wall.

Thus, we see that it will be impossible for the crowd (taken to mean a gas) to give all its energy to a mobile wall.

The fundamental mechanism that leads to the **second principle of thermodynamics** *has just been demonstrated.*

These "light-hearted" images, which will perhaps not please everyone, were an oral support for the presentation of different transport phenomena by one of the authors. We hope that the reader, once they have studied this book, will want to return to this text. They will then have understood, we hope, the images that lead to the development of thermodynamics.

And if this text has a moral, it would be: *Writing down thermodynamics, just like thermal science, is based on continuous equations. The fundamentals of physics that determine these phenomena arise from the field of the discontinuous: discontinuity of matter, divided into particles; discontinuity of light, divided into photons.*

List of Notations

Some notations that are systematically used in this book are defined below.

L	Length
l	Width
H, h	Height
s, S	Surface
e	Thickness
x	Abscissa
y	Ordinate
z	Dimension (third dimension)
T	Temperature
c	Mass fraction
ρ	Mass density
μ	Dynamic viscosity
ν	Kinematic viscosity
c	Calorific mass capacity

λ	Thermal conductivity
a	Thermal diffusivity
D_i	Diffusion coefficient (species i)
Pr	Prandtl number
Sc	Schmidt number
h	Convection coefficient
R_L, R_D, R_e, R_x, etc.	Reynolds number
Nu_L, Nu_x, Nu_D, Nu_e, etc.	Nusselt number
St_x, St_L	Stanton number
Sh_L, SH_x, Sh_D, Sh_e, etc.	Sherwood number
St_{Diffx}, St_{DiffL}	Diffusive Stanton number
Pe	Péclet number
Φ	Thermal flow
φ, φ_W	Thermal flow density
$\dot{m}$	Mass flow
J	Mass flow density
R_{th}	Thermal resistance
k	Thermal resistance (exchangers)
k	Chemical kinetics

1

General Notions

1.1. General notions

The transport of heat from one point to another in space is the result of three fundamental modes: radiation, conduction and convection.

First, let us remember that heat results from the kinetic energy of molecules for a gas or a liquid, or of atoms for a solid. These molecules are indeed, at any temperature, subject to erratic movements and collisions. The atoms of a crystal are more bonded and vibrate around their equilibrium position. Boltzmann related the temperature to the root mean square value of the velocity of these elements in thermal motion. The transfer of this thermal motion from one point to another therefore results from three processes.

Radiation constitutes a singular point: energy is transported in the form of an electromagnetic wave. Here, there is a phenomenology at two scales: the scale of the fluid and the scale of the atom. It is not thermal motion that is transmitted. The shocks resulting from the temperature cause electronic transitions that generate radiation. When this radiation is sent through a solid (surface) or a fluid, this electromagnetic energy is absorbed by the atoms of the receiver, which transform it back into kinetic energy. The molecules or atoms are thermalized. There is absorption.

Thermal motion being the result of shocks, thermal kinetic energy can be transmitted by shocks from a zone of a medium to another zone of the same medium, where the thermal motion has a lower level or at an interface to another medium (gas–solid transfer, liquid–solid transfer, for example). This transfer by shocks is the basis of ***thermal conduction***.

Finally, when a certain amount of heat has been transmitted to a medium, this medium can be moved in space. This is the phenomenon of convection.

Anyone who heats water in a pot and moves that pot to the mug in which they are making their tea is performing ***convection***; many do so without knowing it.

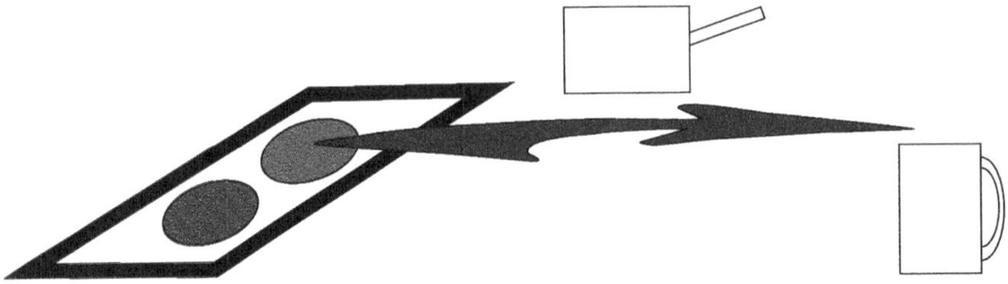

Figure 1.1. *A very basic form of convection. For a color version of this figure, see www.iste.co.uk/ledoux/heat3.zip*

In practice, heat is transferred through a flow, which allows for, among other things, a continuous process. Convection then becomes a chapter of fluid mechanics.

We have two types of problems to solve:

– The generation of the flow and its behavior from the place of "heat collection" to the place of "heat deposition".

– The transfer of heat from a source to the flow. This transfer is generally done at a solid interface. A set of specific phenomena involving coupling between the effects of viscosity and thermal conduction appears. The flow stratifies in speed and temperature, leading to significant thermal gradients perpendicular to the wall. This is called the boundary layer.

The first of these two problems is largely related to "pure" fluid mechanics. We refer to the two works by the same authors devoted to this subject.

Radiation and conduction can be distinguished. We have devoted two previous volumes to them in this series.

The second problem, heat transfer from a source to the carrier fluid, is in fact the main subject of what is called *thermal convection* in the literature. This will be one of the focuses of this work.

This classification, which is necessary for the structuring of books, should never make us lose sight of the fact that the different modes of transfer are, in practice, very often coupled. We have already experienced this in the two previous books devoted respectively to conduction and radiation.

1.2. Forced convection, natural convection

In the exchanges between a panel and a fluid, two fundamental situations can be isolated.

a) The flow of the fluid along the panel is imposed by adequate mechanical means. This is called forced convection.

The flow of the fluid (liquid or gas) can be guided by a conduit: this is internal forced convection. This is also called the internal boundary layer or flow velocity profile.

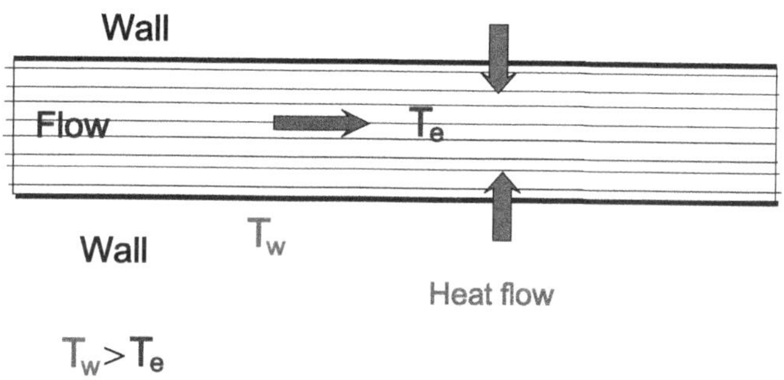

Figure 1.2. *Internal forced convection. For a color version of this figure, see www.iste.co.uk/ledoux/heat3.zip*

The fluid flow can be of indefinite extension and touch the panel. This is called external forced convection. It is also called the external boundary layer. In these problems, the observed velocity profile shows a strong normal gradient to the panel over a "small" thickness, and an asymptotic connection to the "far" flow.

b) Flow can be determined by the temperature difference between the fluid and the wall. For both liquids and gases, the density can vary with temperature. Fluids lighten with temperature (at constant pressure). In the case where temperature gradients lighten the fluid located at the bottom of a device, gravity forces generate a flow. This takes the form of a "convection cell": the fluid "descends" away from the panel and consequently "rises" to the panel. This happens regardless of the inclination between the panel and the horizontal side.

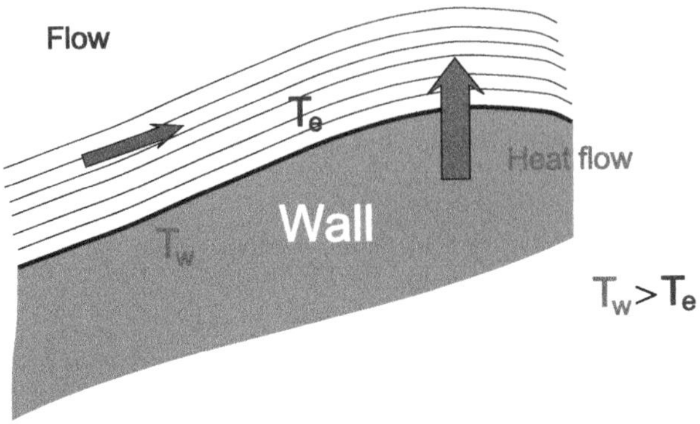

Figure 1.3. *External forced convection. For a color version of this figure, see www.iste.co.uk/ledoux/heat3.zip*

This phenomenon defines natural convection. We sometimes come across the term "free convection".

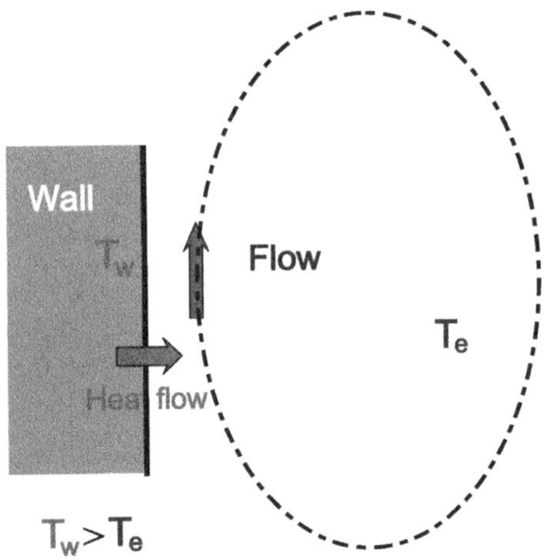

Figure 1.4. *Natural convection. For a color version of this figure, see www.iste.co.uk/ledoux/heat3.zip*

1.3. The calculation of heat transfer

In these two cases of convection, the treatment of a problem can often be carried out in two steps:

a) Determination of the non-transfer flow, in other words, determining the internal velocity field of the fluid in a conduit.

b) Calculation of the temperature field and the parietal flows.

This procedure will be developed further below. The two-step presentation corresponds to many practical cases. Some problems can, unfortunately, be more complex. The properties of the fluid (density, viscosity) can strongly depend on the temperature. As we will see in Chapter **3**, in this case, we have to solve three coupled equations: the continuity equation and momentum equation for the flow, and the energy equation for the temperature field. In most practical problems, the transfer is carried out in the steady state. If this is not the case, we are working in the "input regime". The velocity profile is then variable from one tube section to another.

In many practical cases, we seek to simplify the problem as much as possible. Thus, empirical relations are used to calculate the transfer. These relations can be the result of an experimental study or of a theoretical calculation that has been previously made and published in the literature.

In this type of approach, the convection coefficient becomes an essential element.

1.4. Convection coefficient

For the practical calculation of heat exchange between a solid panel and a flow that flows along this panel, an intermediate calculation called the convection coefficient is used.

Φ is the heat flux (measured in) that is transferred through a panel surface S. T_e is a characteristic temperature of the fluid flow, and T_W is a parietal temperature.

The convection coefficient, which will be noted systematically here as h, is defined by:

$$\Phi = h\,S\left(T_e - T_W\right) \tag{1.1}$$

We can, as we will often do here, get the following in terms of heat flux density $\varphi_W = \frac{\Phi}{S}$. We then clearly have:

$$\varphi_W = h\left(T_e - T_W\right) \qquad [1.2]$$

Note that although the values of Φ or φ_W are chosen to be positive, we have just written the expression of a flow going from the wall to the fluid (heating of the fluid), which implies $T_W - T_e > 0$. In the case of a flow going from the fluid to the panel (heating of the wall), we should write $\Phi = h\,S\left(T_e - T_W\right)$ or $\varphi_W = h\left(T_e - T_W\right)$ with $T_e - T_W > 0$.

Several important points should be considered at this stage:

The definition of T_e can vary from one problem to another: T_e can be an axial temperature in a tube, a temperature "averaged" over a tube section, or the temperature of a fluid far away from the wall.

T_e as well as T_W have no purpose, except specified or assumed to be constants. They can vary with the longitudinal coordinate (noted systematically in the problem). They can vary with the longitudinal coordinate (noted systematically as x of the problem).

In this case, the expression h can be affected by the laws of variation $T_e(x)$ and $T_W(x)$.

This remark is important because, by simplification, expressions of h deduced from theories with constant T_W are often used in variable T_W problems. This is, in particular, systematically the case in the calculations of exchangers that we will discuss in Chapter **4**.

The convection coefficient is a priori a function of different parameters of the problem: physical properties of the fluid (dynamic or thermal) and dynamic characteristics of the flow. Moreover, it is not clearly universally independent of the $T_e - T_W$ temperature difference. The physical properties of the fluid are indeed, in some problems, dependent on the temperatures of the fluid.

In a case where the dynamic and thermal properties of the fluid are constant, the forced convection coefficient will be independent of the $T_e - T_W$ temperature difference.

This will not be the case in natural convection, where, as we will see, h depends on a Grashof number, itself containing the $T_e - T_W$ difference in its definition.

1.5. The program of our study

We have not yet discussed the direction of the transfer between the fluid and wall. In the case of forced convection, we will find a symmetrical problem: the same expression of the convection coefficient will intervene whether the panel is heated or cooled (in the case where the properties of the fluid are constant).

The case of natural convection is evidently asymmetrical; the panel must be heated for there to be transfer.

Using the terms provided in **section 1.3**, we will describe two approaches to the calculation of convective transfer:

a) An "empirical" approach, which will use the convection coefficient, as we will see, as well as a set of dimensionless numbers. This subject will be discussed in Chapter **2**.

b) A theoretical approach, through the boundary layer theory, more complex and also more complete. Within this approach, we will insist on the integral methods, which are approximate, but very useful in practice. This subject will be discussed in Chapter **3**.

On a practical level, we will present some notions on the theory of exchangers. Even though the framework of this book forbids an extensive treatment of the question (let us repeat that we are not writing a treatise, nor a handbook), we feel it is necessary to introduce this theory, which does not shy away from a certain level of approximation, but which still structures the design of classical exchangers, even in an increasingly digitalized space. This subject will be discussed in Chapter **4**.

A final distinction between flows has not yet been introduced here, which also segments the study of convection into two parts of unequal difficulty.

Chapter **2** to Chapter **4** will only deal with single-phase flows, which will remain either liquid or gaseous in their entirety.

In the case of heating a liquid, it can, under certain conditions, be converted into a gas. A flow of steam that is cooled can condense. This is called a two-phase flow. The heat exchange to obtain these phase changes must be consequent. Indeed, a two-phase flow mixes the exchange of sensitive heat and the exchange of latent heat.

NOTE.– Note that it takes 419 kJ to heat a liter of water from 0°C to 100°C, while it takes 2,257 kJ to evaporate this same liter of water to 100°C.

It is then conceivable that the convection coefficients will be significantly increased in the presence of convection with vaporization. The calculation of such flows will become particularly complex, but this type of transfer will be favored whenever a significant "thermal load" is required. This is the case, in particular, in the nuclear steam industry for nuclear power generation.

2

Empirical Approaches

2.1. Introduction

The empirical approach is guided by the assessment of a convection coefficient. This coefficient is often found in the literature in written relations between dimensionless number s or dimensionless criteria (the expression seems to be of French origin or tradition). This practice concerns both experimentally established relations and the results of theoretical approaches.

Convection covers a large number of physical situations differing both by geometries and experimental conditions. The technical interest of the subject is evident and has led, in the preceding decades and still today, to an impressive number of works, both theoretical and experimental. The theoretical approach, based on mathematical analysis, the solution of the equations of fluid mechanics explained below, has for many years been coupled with digital approaches of varying degrees of complexity, which allow a more refined approach to particular problems. In the spirit of this book, we will restrict ourselves to analytical approaches in this chapter.

The very rich literature in the field is, occasionally, the subject of exhaustive summaries of the results available for the calculation of convection coefficients and flows. These works, known as Handbooks, are an essential reference for the professional. Their volume and cost exceed the needs of the student, or even of the technician or engineer who only occasionally approaches thermal calculation. As the latter constitutes the readership we are aiming at, we will restrict ourselves to a few essential relations, referring the reader to specialized briefs for more specific cases (such as for exchangers) or to the consultation of a Handbook in the library. The "Web" is also a source, but its knowledge must be applied with care.

2.2. The dimensionless numbers (or dimensionless criteria) of convection

These dimensionless "criteria" are fundamental in the research or presentation of many results in transfer physics, as in general in fluid mechanics.

2.2.1. *The interest of the dimensionless representation is, at first sight, twofold*

a) Practical interest. We may think that it would be more "practical" to have explicit relations linking the convection coefficient to the data of the problem, fluid characteristics, flow data and possibly temperature data. Such relations are sometimes available. They may come from authors living in different parts of the world. However, it turns out that, by habit or otherwise, many authors remain faithful to the Anglo-Saxon system of units, which we know, even though its use is not the most convenient. One of the authors of this book once came across a formula with two lengths: one in the numerator expressed in inches and the other in the denominator expressed in yards. There is not a small risk of errors in such situations. The dimensionless representation of the results thus makes it possible to complete our calculations without ambiguities.

b) However, there is another motivation, arguably more important.

The dimensionless criteria are the central element of dimensional analysis. This technique is applicable in all of physics, but has known major success in fluid mechanics. Indeed, dimensional analysis is the basis of the model theory, which has made it possible to considerably simplify the studies in wind tunnels, both by easing the experimental plans and reducing the size of the necessary installations. The science of heat transfer, somehow part of fluid mechanics, has followed this movement.

2.2.2. *Vaschy–Buckingham theorem*

This theorem is the basis of the practical application of dimensional analysis. Initially established by the Frenchman Joseph Bertrand, it is universally referred to under the names of the two above-mentioned authors. It is also known as the PI theorem.

It is shown that if a physical law involves n physical variables, and the latter involves k fundamental units, then this law can be put in the form of a relation between $(n-k)$ independent dimensionless parameters, built from the physical variables identified here.

We will not demonstrate this theorem here; nor will we give a general method for determining the construction of the criteria. We will use relations that have already been constructed; the curious reader can consult a fluid mechanics textbook that develops this subject (in particular, Brun, Matthieu, Martinot Lagarde).

2.2.3. *Definition and significance of the dimensionless criteria of fluid mechanics and heat transfer*

2.2.3.1. *Definitions*

Let us repeat that the definition of these numbers is generally the result of the dimensional analysis of the problems encountered.

More than 100 of these criteria can be found in the literature. Some of them only appear rarely, and they are usually the result of a very specific problem.

We will mention the most common ones here. It will be interesting to note that they can very often be interpreted as the ratio of two orders of magnitude of two competing phenomena.

Several types of dimensionless numbers are encountered in convection. These numbers can be:

– attached to the heat transfer fluid;

– attached to the convective flow;

– directly related to the flows, or rather to the density of thermal flows involved.

In Volume 4, Chapter 2, we will see that similarly constructed numbers can be found in boundary layers with mass or chemical species transfer.

A number attached to the fluid is unavoidable, it is the Prandtl number

The ratio of the kinematic viscosity $\nu = \dfrac{\mu}{\rho}$ to the thermal diffusivity $a = \dfrac{\lambda}{\rho c_P}$ of the fluid:

$$\mathrm{Pr} = \frac{\nu}{a} \tag{2.1}$$

Note that this number can also be written from the definitions of ν and a:

$$\Pr = \frac{\mu c_P}{\lambda} \tag{2.2}$$

A number attached to the flow is the Reynolds number

The Reynolds number cannot be ignored as we approach viscous fluid mechanics. Let us recall its definition: it is constructed from a velocity, a length and a kinematic viscosity. The parameters adopted will depend on the problem. The length, for example, can be a distance to the leading edge for a flow along a wall, x, or a diameter, D, for a conduit flow. The corresponding velocities will then be a potential flow velocity ("far" flow), U_e, or a flow velocity in a pipe, V_q.

Generally speaking:

$$R_L = \frac{V\,L}{\nu} \tag{2.3}$$

And we will have, for example:

$$R_x = \frac{U_e\,x}{\nu} \tag{2.4}$$

$$\text{or } R_D = \frac{V_q\,D}{\nu} \tag{2.5}$$

We also define the coefficient of friction:

$$C_f = \frac{2\,\tau_W}{\rho\,U_e^2} \tag{2.6}$$

where $\tau_W = \mu \left.\frac{\partial u}{\partial y}\right|_W$ is the tangential stress (viscosity forces) applied by the fluid on the wall.

We have written it as C_f here in the case of a plate, which will be of interest. We find something similar in a more generalized way in aerodynamics, for example, $C_D = \frac{2\,F}{\rho V^2}$ for the forces of resistance to the motion of a solid.

Several numbers are initially attached to the parietal flow density

The Nusselt number is constructed from the convection coefficient, the thermal conductivity of the heat transfer fluid and a length:

$$Nu = \frac{h\,L}{\lambda} \qquad [2.7]$$

The length can also be an abscissa (distance to the leading edge) in the case of external exchange with a wall, for example. The length can also be a pipe diameter, in case of internal convection. Thus, we often have to consider $Nu_x = \frac{h\,x}{\lambda}$ or $Nu_D = \frac{h\,D}{\lambda}$.

Another number often used in the case of a plate is the Stanton number:

$$St = \frac{h}{\rho\, c_P\, U_e} \qquad [2.8]$$

We will not use this number for the assessment of convection coefficients, favoring the Nusselt in application. However, this number becomes important in the application of integral methods.

In the literature, often old, the Stanton is known as the Margoulis criterion. Note the relationship between Nusselt and Stanton:

$$St = \frac{h}{\rho\, c_P\, U_e} = \frac{h\,D\,\lambda}{\rho\, c_P\, U_e\, D\,\lambda} = \frac{hD}{\lambda}\frac{a}{U_e\,D} = \frac{hD}{\lambda}\frac{\nu}{U_e\,D}\frac{a}{\nu} \qquad [2.9]$$

So, finally:

$$St = Nu_D\; R_D^{-1}\, \mathrm{Pr}^{-1} \qquad [2.10]$$

A less frequently encountered number, but of historical importance, is the Péclet number.

The Nusselt number Nu allows the direct calculation of the convection coefficient.

The Péclet number Pe is less used. However, it is used in some transfer formulas. It is often found in textbooks from the 1950s.

This does not prevent Péclet (1793–1857) from having been a great physicist and engineer in the 19th century, who notably left an important thermal treaty.

We define this number by:

$$Pe = \frac{U_e\, x}{a} \qquad [2.11]$$

Finally, in the problems of natural convection, the Grashof number will appear, defined from the acceleration of gravity, the expansion coefficient of the heat transfer fluid, a characteristic length of the phenomenon and the kinematic viscosity of the fluid. All of these parameters are naturally present (we develop a boundary layer from a flow). However, in this case, the flow is also determined by the temperature (since the engine is gravity intervening in a differential way on a fluid which expands). A difference in temperature will also be observed, generally between the wall and the fluid (stationary) at the distance: $\Delta T = T_W - T_e$. This will introduce a nonlinearity between the parietal flow density and the temperature difference.

$$Gr = \frac{g\, \beta\, L^3\, \Delta T}{\nu^2} \qquad [2.12]$$

The expansion coefficient is defined by:

$$\beta = \frac{1}{V_{OL}} \frac{dV_{OL}}{dT}\bigg|_p = -\frac{1}{\rho}\frac{d\rho}{dT}\bigg|_p . \qquad [2.13]$$

It will vary depending on the fluid. For a perfect gas, we have:

$$\beta = \frac{1}{T} \qquad [2.14]$$

this is often the case for any gas (which is valid as long as the pressures remain low).

For a vertical flat plate, a local Grashof will be used and L will be an abscissa x.

2.2.3.2. *Interpretation*

The dimensionless criteria are not simple mathematical constructions resulting from the dimensional analysis, they are susceptible to a physical interpretation, generally based on the relationship of order of magnitude between two phenomena.

a) The Prandtl number is the ratio of the kinematic viscosity and the thermal diffusivity.

These two parameters are related to dynamic phenomena: thermal diffusivity is involved in unsteady heat conduction problems and kinematic viscosity can also be interpreted in terms of a dynamic parameter in the unsteady evolution of a velocity field. This aspect is dealt with in a work by the same authors of this book, devoted to fluid mechanics [LED 17b].

b) The Reynolds number, $R_x = \dfrac{U_e x}{\nu}$, can be rewritten as:

$$R_L = \frac{U_e L}{\nu} = \frac{\rho U_e L}{\mu} = \frac{\rho U_e^2}{\mu \dfrac{U_e}{L}} \qquad [2.15]$$

After this slight manipulation, it is interpreted as the ratio of a pressure (flow of momentum) to a tangential stress. It compares the energy contained in the kinetics of the flow to the braking by the wall. Less than an energy assessment, we also measure the stability of the flow in question.

To interpret the Nusselt, we will place ourselves, for example, in the case of a pipe and introduce the temperature difference between the axis and the wall of this tube $T_e - T_W$:

$$Nu_D = \frac{h D}{\lambda} = \frac{h\left(T_e - T_W\right)}{\lambda \dfrac{T_e - T_W}{D}} = \frac{\varphi_W}{\lambda \dfrac{T_e - T_W}{D}} \qquad [2.16]$$

The Nusselt then appears as the ratio of the parietal flow density and a flow driven by conduction if the fluid was stationary. We thus compare convection and pure conduction.

The Stanton draws another comparison. We see this by once again introducing the difference $T_e - T_W$:

$$St = \frac{h}{\rho c_P U_e} = \frac{h(T_e - T_W)}{\rho c_P U_e (T_e - T_W)} = \frac{\varphi_W}{\rho c_P U_e (T_e - T_W)} \quad [2.17]$$

We can see the ratio between the flow density (flow per m²) of “sensible” heat transported by the flow, $\rho c_P U_e (T_e - T_W)$ and the parietal flow density “collected” by convection, φ_W. As some authors have written, the Stanton measures a “sweeping capacity” of the flow in terms of heat.

The ***Péclet*** can be rewritten as:

$$P_e = \frac{x^2}{a} \frac{U}{x} \quad [2.18]$$

We can see the double relationship here:

$$\begin{aligned} \tau_{transport} &= \frac{U}{x} \\ \tau_{conduction} &= \frac{x^2}{a} \end{aligned} \quad [2.19]$$

The transport time $\tau_{transport}$ is the time it takes for the flow to travel distance x.

The conduction time $\tau_{conduction}$ comes from a formula given by Einstein (established for the Brownian motion):

$$x^2 = 2at \quad [2.20]$$

where x is the order of magnitude of the distance reached by an unsteady “conduction wave” at time t.

To understand this formula, remember the definition of η in the problems of unsteady conduction:

$$\eta = \frac{x}{\sqrt{4at}} \quad [2.21]$$

Péclet will therefore be the ratio of transport time to conduction time.

$$P_e = \frac{x^2}{a}\frac{U}{x} = \frac{\tau_{transport}}{\tau_{conduction}} \quad [2.22]$$

We also define, by analogy, a diffusive Péclet number, $\frac{V\,x}{D}$ in mass transfer.

2.3. Calculation of convection coefficients: external convection

We will mainly express these coefficients from the Nusselt numbers.

The formulas given in the following assume, as a rule, that the wall temperature T_W is homogeneous. In the case of tubes, this assumption can be accepted in many cases, in the case of plates "heading upwind", this assumption is less universal and significant errors may occur.

We will also have to distinguish the laminar and turbulent flow regimes each time.

2.3.1. *Case of a flat plate at constant temperature*

2.3.1.1. *First, we must define the flow regime of the fluid*

We know that we distinguish between laminar and turbulent flow. We define a Reynolds number based on the abscissa x, in other words, the distance from a point on the plate to the leading edge.

$$R_x = \frac{U_e\,x}{\nu} \quad [2.23]$$

The thickness of the boundary layer increases with this abscissa, the flow will be destabilized downstream and a transition will appear. The laminar regime will thus be laminar for small Reynolds, and turbulent for its larger values, this being on the same plate.

The following criteria are adopted:

– The regime is considered to be laminar for $R_x < 5.10^5$.

– The regime is considered to be turbulent for $R_x > 5.10^5$.

There is therefore, theoretically, a laminar transfer zone that precedes a turbulent zone to "infinity". In practice, this is not a real problem. As the Reynolds numbers are generally of strong powers of 10, the contribution to the transfer of the laminar zone is negligible when the regime is turbulent.

2.3.1.2. *Calculation in laminar regime*

The formulas given here result from the theories of Blasius (for the dynamic part) and Polhausen (for the thermal part).

Let us recall in passing the values of the various characteristic thicknesses of the boundary layer:

Boundary layer thickness (defined by the ordinate where $\frac{u}{U_e}$ = 0.99):

$$\frac{\delta}{x} = \frac{4.92}{\sqrt{R_x}} \qquad [2.24]$$

Integrated dynamic thicknesses (integral methods can be seen later):

$$\frac{\delta_1}{x} = \frac{1.72}{\sqrt{R_x}} \qquad [2.25]$$

$$\frac{\delta_2}{x} = \frac{0.664}{\sqrt{R_x}} \qquad [2.26]$$

The local friction coefficient is: $C_f = \frac{0.664}{\sqrt{R_x}}$.

Nusselt numbers

The Nusselt number can be assessed locally, in other words, at a given abscissa x. We then have:

$$Nu_x = 0.332\, R_x^{0,5}\, P_r^{1/3} \qquad [2.27]$$

We can also define an average transfer coefficient (Nusselt) over a length L of a plate, starting from the leading edge. It is defined from the average flow over this length. If l is the width of the plate:

$$\bar{\Phi} = \frac{1}{l\,L}\int_0^L \varphi_W(x)\, l\, dx \qquad [2.28]$$

and

$$\bar{\Phi} = \bar{h}\, lL\,(T_e - T_W) = h_L\, lL\,(T_e - T_W) \qquad [2.29]$$

Evidently:

$$h_L = \frac{1}{L}\int_0^L h(x)\; dx = \frac{1}{L}\int_0^L \frac{\lambda\, Nu_x}{x}\; dx = \frac{1}{L}\int_0^L \frac{\lambda}{x} 0.332\sqrt{\frac{U_e}{\nu}}\; x^{0.5}\; \mathrm{Pr}^{1/3}\, dx \qquad [2.30]$$

We will note two equivalent notations for the average convection coefficient, $\bar{h}$ or h_L.

Note that we will use $(T_e - T_W)$ or $(T_W - T_e)$ according to the sign of the transfer (heating or cooling of the plate):

$$Nu_L = \frac{h_L\, L}{\lambda} = \left[\frac{1}{L}\int_0^L h(x)\; dx\right]\frac{L}{\lambda} = \left[\frac{1}{L}\int_0^L \frac{\lambda\, Nu(x)}{x}\; dx\right]\frac{L}{\lambda}$$

$$= \int_0^L 0.332\sqrt{\frac{U_e}{\nu}}\;\; x^{-0.5}\; dx = \frac{0.332}{0.5}\sqrt{\frac{U_e}{\nu}}\;\; x^{0.5}\Big|_0^L \qquad [2.31]$$

Therefore:

$$Nu_L = 0.664\sqrt{\frac{U_e L}{\nu}}\; \mathrm{Pr}^{1/3} \qquad [2.32]$$

$$Nu_L = 0.664\; R_L^{0.5}\; \mathrm{Pr}^{1/3} \qquad [2.33]$$

Stanton numbers

We can deduce the local and average Stanton numbers:

$$St_x = 0.332\; R_x^{-0.5}\;\; P_r^{-2/3} \qquad [2.34]$$

$$St_L = 0.664\, R_L^{-0.5}\, \mathrm{Pr}^{-2/3} \qquad [2.35]$$

Note the relationship of Stanton to the friction coefficient:

$$C_{fx} = \frac{0.664}{\sqrt{R_x}} = 0.664\, R_x^{-0.5} \qquad [2.36]$$

$$St_x = \frac{C_{fx}}{2} P_r^{-2/3} \qquad [2.37]$$

This proportionality is not surprising; as highlighted above, these two coefficients express sweeping capacities. The proportionality coefficient expresses the relation of the transport of momentum and heat.

2.3.1.3. *Calculation in turbulent regime*

The mean velocity profile in a turbulent regime is complex. We sometimes use an approximate profile shape (1/7 profile):

$$\frac{u}{U_e} = \left(\frac{y}{\delta}\right)^{\frac{1}{7}} \qquad [2.38]$$

We give the characteristic thicknesses

$$\frac{\delta_1}{x} = \frac{0.046}{R_x^{0.2}} \qquad [2.39]$$

$$\frac{u}{U_e} = 0.735 \left(\frac{y}{\delta_1}\right)^{\frac{1}{7}} \qquad [2.40]$$

Several values for the friction coefficient have been proposed, mainly including:

$$C_f = \frac{0.0592}{R_x^{0,2}} \qquad [2.41]$$

or

$$C_f = \frac{0.0456}{R_x^{0.25}} \qquad [2.42]$$

We will adopt the following as local transfer coefficients (Nusselt)

$$Nu_x = 0.0288\, R_x^{0.8}\, \mathrm{Pr}^{1/3} \qquad [2.43]$$

We deduce the average transfer coefficient (Nusselt) on L.

The local convection coefficient is equal to:

$$h_x = \frac{\lambda Nu_x}{x} = 0.0288\, \lambda \left(\frac{U_e}{\nu}\right)^{0.8} x^{-0.2}\, \mathrm{Pr}^{1/3} \qquad [2.44]$$

$$Nu_L = \frac{L}{\lambda}\frac{1}{L}\int_0^L h_x\, dx = \frac{L}{\lambda}\frac{1}{L}\int_0^L 0.0288 \left(\frac{U_e}{\nu}\right)^{0.8} P_r^{\frac{1}{3}}\, x^{-0.2} dx \qquad [2.45]$$

$$Nu_L = 0.0288 \left(\frac{U_e}{\nu}\right)^{0.8} \mathrm{Pr}^{1/3} \left.\frac{x^{0.8}}{0.8}\right|_0^L \qquad [2.46]$$

$$Nu_L = \frac{0.0288}{0.8} \left(\frac{U_e L}{\nu}\right)^{0.8} \mathrm{Pr}^{1/3} \qquad [2.47]$$

$$Nu_L = 0.036\, R_L^{0.8}\, \mathrm{Pr}^{1/3} \qquad [2.48]$$

Stanton numbers

We will deduce the Stanton numbers:

$$St_x = 0.0288\, R_x^{-0.2}\, \mathrm{Pr}^{-2/3} \qquad [2.49]$$

$$St_L = 0.036\, R_L^{-0.2}\, \mathrm{Pr}^{-2/3} \qquad [2.50]$$

And we will note that, like in the laminar regime:

$$C_{fx} = 0.0592\, R_x^{-0.2} \qquad [2.51]$$

$$St_x = \frac{C_{fx}}{2} P_r^{-2/3} \qquad [2.52]$$

2.3.2. *External convection on an obstacle: case of a tube outside a flow*

This practical case is particularly encountered at two scales: in the measurement of velocities by hot wire (the wire would be of very small diameter, sometimes of the order of a few microns) and in the tube exchangers.

The Reynolds and Nusselt numbers will be based on the tube diameter.

$$R_D = \frac{V\,D}{\nu} \quad [2.53]$$

where V is the wind velocity impacting the tube.

$$Nu_D = \frac{h\,D}{\lambda} \quad [2.54]$$

The Nusselt will then be formulated as:

$$Nu_D = C\,R_D^n\,\mathrm{Pr}^{1/3} \quad [2.55]$$

The coefficients C and n will vary with the Reynolds number, according to Table 2.1.

R_D	C	n
0.4 – 4	0.98	0.33
4 – 4	0.911	0.385
40 – 4000	0.83	0.466
4000 – 40 000	0.193	0.618
40 000 – 250 000	0.0266	0.805

Table 2.1. *Coefficients C and n*

2.4. Internal convection

2.4.1. *Convection in a tube*

This case is the most frequently encountered.

2.4.1.1. *The Reynolds and Nusselt numbers will be based on the tube diameter*

$$R_D = \frac{V_q D}{\nu} \qquad [2.56]$$

where V is the wind velocity impacting the tube.

$$Nu_D = \frac{h D}{\lambda} \qquad [2.57]$$

Here, V_q is the flow velocity in the tube. It is deduced from the mass flow rate, or the volume flow rate of the flow. Physically, V_q is the velocity that the flow would have if it were uniform in the tube. In other words, it is the velocity of the perfect fluid flow that would have the same flow rate as the actual fluid flow treated here.

$$V_q = \frac{4 Q_V}{\pi D^2} = \frac{4 Q_m}{\rho \pi D^2} \qquad [2.58]$$

Q_V and Q_m are respectively the volume and mass flow rates of the flow.

2.4.1.2. *A particular element comes into play here: the length of establishment of the regime*

The traditional formulas given below are valid in the case of an established regime, that is, a velocity profile identical from one section to another of the pipe. Indeed, when a uniform flow is introduced into a pipe, a form of boundary layer develops in the vicinity of the pipe origin. The growth of this boundary layer leads to an established flow, on a length of the so-called establishment of the dynamic regime. The same problem arises for the temperature profiles and a length of establishment of thermal regime will then be defined. To correctly apply the formulas given here, it is necessary to ensure that the useful length of the tube is significantly greater than the dynamic and thermal establishment lengths.

We will, of course, also have to distinguish the laminar flow regime from the turbulent flow regime.

The following relationships are generally adopted to assess a dynamic establishment length L_e and a thermal establishment length L_{th}; these lengths are calculated with reference to the tube diameter:

For the ***dynamic regime***

$$\frac{L_e}{D} = 0.05\ R_D \quad [2.59]$$

For the ***thermal regime***

$$\frac{L_{th}}{D} = 0.05\ R_D\ \Pr \quad [2.60]$$

Critical Reynolds: 2,200

2.4.1.3. *Laminar regime*

Let us recall that this flow regime in a tube is generally accepted for a Reynolds number value that is less than $R_D \approx 2,200$. This criterion may vary according to the authors. In practice, industrial flows are turbulent, Reynolds are much higher, and thoughts on the criterion in laminar flow remain theoretical.

The result given here results from a theory established by Graetz. In the established laminar regime, for a constant wall temperature, the Nusselt based on the dimeter is a constant:

$$Nu_D = 3.66 \quad [2.61]$$

2.4.1.4. *Turbulent regime*

The results here are of experimental origin. The *Colburn formula* can be used:

$$Nu_D \quad = 0.023\, R_D^{0,8}\ P_r^{1/3} \quad [2.62]$$

In an unestablished regime, some empirical expressions have been proposed, including the following:

$$Nu(x) = 0.06\left(\frac{2 \quad x}{D\, R_D\ \Pr}\right)^{1/3} \quad [2.63]$$

2.4.2. *Forced convection between two plates*

This flow geometry is found in the field of heat exchangers (plate heat exchangers).

2.4.2.1. *The flow is bounded by two plates at a distance from e*

This distance will be the reference length for the construction of the Reynolds and Nusselt:

$$R_e = \frac{V_q\, e}{\nu} \qquad [2.64]$$

where V_q is the flow velocity.

$$Nu_e = \frac{h\, e}{\lambda} \qquad [2.65]$$

We note that the results are formally close to what we find in pipe flow. Nevertheless, the validity of these formulas remains conditioned to the form factor of the flow: the plates must remain very wide in relation to their distance.

We will use the same Reynolds criteria, based on the inter-plate distance, to differentiate laminar and turbulent regimes.

2.4.2.2. *Laminar regime*

In an established regime, the Nusselt is constant:

$$Nu_e = 3.4 \qquad [2.66]$$

2.4.2.3. *Turbulent regime*

We will use Colburn's formula:

$$Nu_e = 0.023\, R_e^{0.8}\, \mathrm{Pr}^{1/3} \qquad [2.67]$$

2.5. Natural convection

2.5.1. *Let us recall useful dimensionless numbers*

Prandtl number:

$$P_r = \frac{\nu}{a} \qquad [2.68]$$

Grashof number:

$$G_r = \frac{g\,\beta\,L^3\,\Delta T}{\nu^2} \quad [2.69]$$

Expansion coefficient:

$$\beta = -\frac{1}{\rho}\frac{d\rho}{dT} \quad [2.70]$$

which, for a perfect gas, will be reduced to: $\beta = \frac{1}{T}$.

The Nusselt number is defined from a characteristic dimension, noted as L:

$$Nu = \frac{h\,L}{\lambda} \quad [2.71]$$

2.5.2. *Nusselt calculation*

A distinction must be made between laminar and turbulent convection.

However, for all fluids and all regimes, a single generic form is used to calculate the Nusselt:

$$Nu_L = C\left(Gr\,\mathrm{Pr}\right)^n \quad [2.72]$$

The parameter will depend on the flow regime:

– in laminar regime $n = \frac{1}{4}$;

– in turbulent regime $n = \frac{1}{3}$.

For each plan, the parameter C depends on the value of the product $Gr\,\mathrm{Pr}$.

Tables 2.2–2.5 give the values for different geometries. The reference length is specified.

2.5.2.1. *Vertical plate*

The reference length is the height of the plate, in other words, H (see Table 2.2).

Laminar		**Turbulent**	
Range of $Gr\,\mathrm{Pr}$	C	Range of $Gr\,\mathrm{Pr}$	C
$10^4 - 10^9$	0.59	$10^9 - 10^{13}$	0.13

Table 2.2. *C values for different geometries*

2.5.2.2. *Horizontal plate*

The reference length is the width of the plate, in other words, L, in the case where the transfer is upwards (see Table 2.3).

Laminar		**Turbulent**	
Range of $Gr\,\mathrm{Pr}$	C	Range of $Gr\,\mathrm{Pr}$	C
$10^5 - 2.10^7$	0.54	$2.10^7 - 3.10^{10}$	0.14

Table 2.3. *C values for different geometries*

2.5.2.3. *Vertical plate*

The reference length is the height of the plate, in other words, H, in the case where the transfer is downwards (see Table 2.4).

Laminar		**Turbulent**	
Range of $Gr\,\mathrm{Pr}$	C	Range of $Gr\,\mathrm{Pr}$	C
$3.10^5 - 3.10^{10}$	0.23	$3.10^{10} - 10^{13}$	0.07

Table 2.4. *C values for different geometries*

2.5.2.4. *Horizontal cylinder*

The reference length is the diameter of the tube, in other words, D (see Table 2.5).

Laminar		**Turbulent**	
Range of $Gr\,\mathrm{Pr}$	C	Range of $Gr\,\mathrm{Pr}$	C
$10^3 - 10^9$	0.53		0.10

Table 2.5. *C values for different geometries*

2.6. Use of "standard" formulas

In building design, different expressions are prescribed in technical documents. We propose here two formulas to calculate the internal and external convection coefficients.

Interior vertical wall:

$$h_i = 9.09\,W\,m^{-2}\,K^{-1} \qquad [2.73]$$

$$R = \frac{1}{h_i} = 0.11\,m^2 K\,W^{-1} \qquad [2.74]$$

Exterior vertical wall (increase h in case of high wind):

$$h_e = 16.7\,W\,m^{-2}\,K^{-1} \qquad [2.75]$$

$$R = \frac{1}{h_e} = 6.10^{-2}\,m^2 K\,W^{-1} \qquad [2.76]$$

2.7. Some examples of applications

Preparatory work

For all of the following exercises, we collect a set of physical data that may be useful.

Physical properties of air

Taken at $T_a = 20°C$:

Density $\rho = 1.2\ kg.m^{-3}$

Dynamic viscosity $\mu = 1.81.10^{-5}\ Pl$

Thermal conductivity $\lambda = 0.0257\ Wm^{-1}K^{-1}$

Mass heat capacity at constant pressure $c_P = 1006\ J\,kg^{-1}$

From this, we can deduce:

Kinematic viscosity $\nu = \dfrac{\mu}{\rho} = \dfrac{1.81.10^{-5}}{1.2} = 1.5.10^{-5}\ m^2\,s^{-1}$

Thermal diffusivity $a = \dfrac{\lambda}{\rho\, c_P} = \dfrac{0.0257}{1.2*1006} = 2.13.10^{-5}\, m^2\, s^{-1}$

Prandtl number $\Pr = \dfrac{\nu}{a} = \dfrac{1.5.10^{-5}}{2.13.10^{-5}} = 0.7$

Taken at $T_a = 80°C$:

Density $\rho = 0.999\ kg.m^{-3}$

Dynamic viscosity $\mu = 2.09.10^{-5}\ Pl$

Thermal conductivity $\lambda = 0.0302\ Wm^{-1}K^{-1}$

Mass heat capacity at constant pressure $c_P = 1010\ J\,kg^{-1}$

From this, we can deduce:

Kinematic viscosity $\nu = \dfrac{\mu}{\rho} = \dfrac{2.09.10^{-5}}{0.99} = 2.09.10^{-5}\ m^2\,s^{-1}$

Thermal diffusivity $a = \dfrac{\lambda}{\rho c_P} = \dfrac{0.0302}{0.999*1010} = 2.99.10^{-5}\, m^2\, s^{-1}$

Prandtl number $\Pr = \dfrac{\nu}{a} = \dfrac{2.09.10^{-5}}{2.99.10^{-5}} = 0.698 \approx 0,7$

Thermal diffusivity $a = \dfrac{\lambda}{\rho c_P} = \dfrac{0.0359}{0.815*1019} = 4.32.10^{-5}\, m^2\, s^{-1}$

Prandtl number $\Pr = \dfrac{\nu}{a} = \dfrac{2.96.10^{-5}}{4.32.10^{-5}} = 0.685 \approx 0.7$

Physical properties of water

Taken at $T_a = 20°C$:

Density $\rho = 1000\, kg.m^{-3}$

Dynamic viscosity $\mu = 1.10^{-3}\, Pl$

Thermal conductivity $\lambda = 0.597\, Wm^{-1}K^{-1}$

Mass heat capacity at constant pressure $c_P = 4182\, J\, kg^{-1}$

From this, we can deduce:

Kinematic viscosity $\nu = \dfrac{\mu}{\rho} = \dfrac{1.10^{-3}}{1000} = 1.10^{-6}\, m^2\, s^{-1}$

Thermal diffusivity $a = \dfrac{\lambda}{\rho c_P} = \dfrac{0.597}{1000*4182} = 1.43.10^{-7}\, m^2\, s^{-1}$

Prandtl number $\Pr = \dfrac{\nu}{a} = \dfrac{1.10^{-6}}{1.43.10^{-7}} = 6.99 \approx 7$

Taken at $T_a = 80°C$:

Density $\rho = 974\, kg.m^{-3}$

Dynamic viscosity $\mu = 3.55.10^{-4}\ Pl$

Thermal conductivity $\lambda = 0.668\ Wm^{-1}K^{-1}$

Mass heat capacity at constant pressure $c_P = 4196\ J\,kg^{-1}$

From this, we can deduce:

Kinematic viscosity $\nu = \dfrac{\mu}{\rho} = \dfrac{3.55.10^{-4}}{974} = 3.64.10^{-7}\ m^2\,s^{-1}$

Thermal diffusivity $a = \dfrac{\lambda}{\rho c_P} = \dfrac{0.668}{974*4196} = 1.63.10^{-7}\,m^2\ s^{-1}$

Prandtl number $\Pr = \dfrac{\nu}{a} = \dfrac{3.64.10^{-7}}{1.63.10^{-7}} = 2.33$

EXAMPLE 2.1.– Heat flow on a flat plate

A flat steel plate of length $L = 2\,m$ and width $l = 25\ cm$ is maintained at a temperature $T_W = 160\,°\text{C}$. It is swept along its length by a flow of air at temperature $T_e = 20\,°\text{C}$. The distanced velocity is $U_e = 15\,cm\,s^{-1}$.

It is assumed that the transfer between the plate and the air is essentially due to forced convection.

This plate is a gray body of emissivity $\varepsilon = 0.25$. We recall the Stefan constant $\sigma = 5.67.10^{-8}\ J\,m^{-2}\ s^{-1}\ K^{-4}$.

1) What are the local flow densities at the following distances from the leading edge: $x_1 = 4\ cm$, $x_2 = 20\ cm$, $x_3 = 50\ cm$, $x_4 = L$?

2) What is the total flow exchanged between the plate and the air flow?

3) Is the assumption made about the reduction of the transfer to forced convection valid?

4) Repeat the exercise for an air velocity $U_e = 10\,m\,s^{-1}$.

SOLUTION TO EXAMPLE 2.1.–

1) We must first determine the flow regime. The Reynolds number at the end of the plate is:

$$R_L = \frac{U_e L}{\nu} = \frac{0.15 * 2}{1.5.10^{-5}} = 2.10^4$$

A laminar regime will be acknowledged.

We have:

$$Nu_x = 0.332\ R_x^{0.5}\ P_r^{1/3}$$

This gives the following convection coefficient:

$$h = 0.332 \sqrt{\frac{U_e}{\nu}}\ P_r^{1/3} \frac{\lambda}{\sqrt{x}} = 0.332 \sqrt{\frac{0.15}{1.5.10^{-5}}}\ 0.7_r^{1/3} \frac{0.0257}{\sqrt{x}} = \frac{0.76}{\sqrt{x}}$$

Table 2.6 shows the local flow densities $\varphi_W = h\left(T_W - T\right)_e = 140\,h$.

x	$h\ W\,m^{-2}\,K^{-1}$	$\varphi_W\ \ W\,m^{-2}$
$x = 4\ cm$	3.8	532
$x = 20\ cm$	1.7	238
$x = 50\ cm$	1.07	149
$x = L = 2\ m$	0.537	75.2

Table 2.6. *C values for different geometries*

2) The average Nusselt will have the value

$$Nu_L = 0.664\ R_L^{0.5}\ \mathrm{Pr}^{1/3} = 0.664 \sqrt{\frac{0.15 * 2}{1.5.10^{-5}}}\ 0.7^{1/3} = 83.4$$

This gives an average convection coefficient:

$$h_L = \frac{\lambda Nu_L}{L} = \frac{0.0257 * 83.4}{2} = 1.07\ W\,m^{-2}\,K^{-1}$$

And the total convective flux taken at the plate:

$$\Phi = S\,\varphi_W = Ll\,h_L\left(T_W - T\right)_e = 2*0.25*1.07*140$$

$$\Phi = 74.9\,W$$

3) To accept the hypothesis of a predominantly convective transfer, we need to assess the radiative and natural convection exchanges.

3.1) The balance of radiative exchange between the plate and the atmosphere (see parent book on radiation) is:

$$\Phi_R = \varepsilon\,S\,\sigma\left(T_W^4 - T_e^4\right)$$

Remembering that temperatures must be expressed here in Kelvins, this gives:

$$\Phi_R = \varepsilon\,Ll\,\sigma\left(T_W^4 - T_e^4\right) = 0.25*2*0.25*5.67.10^{-8}\left(433^4 - 293^4\right)$$

$$\Phi_R = 197\,W$$

3.2) Transfer by natural convection must be assessed in the absence of a flow.

Horizontal plate: the reference length is the width of the plate, in other words, l.

In this case, the transfer is upwards.

The reference length, in this case, will be the width of the plate.

Let us calculate the Grashof number:

$$G_r = \frac{g\,\beta\,L^3\,\Delta T}{\nu^2}$$

With the air being at $T_e = 20°\text{C}$, we will make an assessment in the following, using the absolute temperature:

$$\beta = \frac{1}{T_e} = \frac{1}{293}$$

This gives:

$$G_r = \frac{g\beta l^3 \Delta T}{\nu^2} = \frac{9.81*0.25^3*140}{293*\left(1.5.10^{-5}\right)^2} = 3.25.10^8$$

The grouping becomes:

$$G_r \Pr = 3.25.10^8 * 0.7 = 2.27.10^8$$

We are in a turbulent regime. The Nusselt is written as:

$$Nu_L = 0.14\left(Gr \Pr\right)^{1/3} = 85.4$$

The exchanged flow will be:

$$\Phi = S\,h\left(T_W - T_e\right) = Ll\frac{\lambda Nu_L}{L}\left(T_W - T_e\right) = 0.25*0.025*785.4*140$$

$$\Phi = 76.8\,W$$

3.3) In conclusion, it can be seen that the convective flow $\Phi = 74.9\,W$ is of the same order of magnitude as the radiative flow $\Phi_R = 197\,W$ and the natural convection flow $\Phi = 76.8\,W$.

This illustrates a situation where we must be very careful in terms of the assumptions made.

Let us note that, with respect to the radiative transfer, the convective calculation could remain valid; indeed, the two transfers are of independent mechanisms. In the event of a thermal balance of the plate, it would obviously be necessary to take the radiative transfer into account.

On the contrary, the calculation made for natural convection invalidates our assessment of the convective transfer.

In fact, as soon as we are in the presence of low-velocity winds (in other words, in a laminar regime), we should carefully test our assumptions.

4) For a velocity $U_e = 10\,m\,s^{-1}$, the Reynolds becomes:

$$R_L = \frac{U_e L}{\nu} = \frac{10*2}{1.5.10^{-5}} = 1.33.10^6$$

We are in a turbulent regime.

We have the local Nusselt:

$$Nu_x = 0.0288\, R_x^{0.8}\, \Pr^{1/3}$$

This gives a convection coefficient:

$$h = 0.0288\left(\frac{U_e}{\nu}\right)^{0.8} P_r^{1/3}\, \frac{\lambda}{x^{0.2}} = 0.0288\left(\frac{10}{1.5.10^{-5}}\right)^{0,8} 0.7^{1/3}\, \frac{0.0257}{x^{0.2}} = \frac{30}{x^{0.2}}$$

Here is the table of local flow densities $\varphi_W = h\left(T_W - T\right)_e = 140\,h$.

x	$h\ W\,m^{-2}\,K^{-1}$	$\varphi_W\ W\,m^{-2}$
$x = 4\ cm$	57.06	7989
$x = 20\ cm$	41.36	5790
$x = 50\ cm$	34.43	4821
$x = L = 2\ m$	26.1	3654

Table 2.7. *C values for different geometries*

2) The average Nusselt will have the value:

$$Nu_L = 0.036\, R_L^{0.8}\, \Pr^{1/3} = 0.036*\left(1.33.10^6\right)^{0.8} *\left(0.7\right)^{1/3} = 2534$$

This gives an average convection coefficient:

$$h_L = \frac{\lambda Nu_L}{L} = \frac{0.0257*2534}{2} = 325.6\ W\,m^{-2}\,K^{-1}$$

And the total convective flow taken at the plate:

$$\Phi = S\,\varphi_W = Ll\,h\left(T_W - T\right)_e = 2*0.25*325.6*140$$

$$\Phi = 22.79\ kW$$

It can be seen that the radiative flows and the natural convection become negligible compared to the forced convection.

EXAMPLE 2.2.– Some principles of wire anemometry

1) Hot wire anemometer

For many years, hot wire anemometry has been the only efficient wind tunnel velocity technique, dedicated in large part to turbulence studies.

These anemometers are based on the thermal exchanges between a thin wire of about $1mm$ in length, and a diameter of 1 μm to 10 μm, stretched between two pins. A simplified model is given here. The metrology is based on the variation of the resistance of the platinum wire with temperature, which allows the measurement.

In the present model, we work at a constant wire temperature and we will take the platinum resistivity at $\rho_{st} = 111.10^{-9}\ \Omega m$.

1.1) Simple hot wire anemometer

A platinum diameter wire $d = 10\ \mu m$ is normally placed at an air wind speed $V = 10\ m\,s^{-1}$ at temperature $T_e = 20°C$.

1.1) What current must be passed through the wire to maintain it at temperature $T_W = 210°C$? It is assumed that the transfer is exclusively convective.

1.2) What would be the parietal flow density exchanged with air in the absence of flow?

2) Impulse anemometer

In the seventies, a French researcher, P. Calvet, proposed a thermal anemometer based on an impulse transfer: short current pulses abruptly increase the temperature of a wire (the initial version used a thermistor), and the impulse response (return to the initial temperature) makes it possible to determine speed and temperature of the flow (see: [TRI 72]).

A simple model of its working principle is given here.

We take the wire in Question **1**, and we suddenly raise its temperature, by a short pulse of intensity, to a temperature $T_{W0} > T_W$.

Show that the return to temperature T_W is exponential and calculate its time constant τ. What frequencies can we expect to measure in turbulence with this wire?

The following properties of the platinum are given:

Resistivity: $\rho_{st} = 111.10^{-9}\ \Omega\, m^{-1}$.

Density: $\rho = 21400\ kg\, m^{-3}$.

Mass heat capacity: $c = 140\ J\, kg^{-1}$.

SOLUTION TO EXAMPLE 2.2.–

1) Wire anemometry

1.1) Or the desired current, $i\ A$.

The resistance per meter R_{1m} of wire is:

$$R_{1m} = \rho_{st}\,\frac{l}{S} = \rho_{st}\,\frac{4}{\pi D^2} = 111.10^{-9}\,\frac{4}{\pi 10^{-10}} = 1413.10^{8}\ \Omega\, m^{-1}$$

The flow per meter of wire and the corresponding flow density are deduced:

$$\varphi_W = \frac{R_{1m}\, i^2}{\pi D} = \frac{1413 i^2}{\pi 10^{-5}} = 4.49.10^{7}\ i^2$$

Let us calculate the convection coefficient:

$$Nu_D = C\ R_D^n\ \Pr^{1/3}$$

The Reynolds number allows us to determine C and n:

$$R_D = \frac{V\,D}{\nu} = \frac{10*10^{-5}}{1.5.10^{-5}} = 6.67$$

We can then determine the temperature difference:

$$C = 0.911$$

$$n = 0.385$$

We then get: $Nu_D = C\,R_D^n\,\Pr^{1/3} = 0.911*6.67^{0.385}*0.7^{1/3} = 1.68$

$$h = \frac{\lambda Nu_D}{D} = \frac{0.0257*1.68}{10^{-5}} = 4318\,W\,m^{-2}\,K^{-1}$$

We must have:

$$\varphi_W = h\left(T_W - T_e\right)$$

Or:

$$4.49.10^{12}\,i^2 = 4318\left(210-20\right)$$

The intensity is derived from the following:

$$i^2 = \frac{4318\left(210-20\right)}{4.49.10^{12}} = 1.83.10^{-2}\ A^2$$

$$i = 0.135\ A = 135\,mA$$

1.2) Let us calculate the flow density by natural convection.

The Grashof number stands at:

$$G_r = \frac{g\,\beta\,L^3\,\Delta T}{\nu^2} = \frac{9.81*10^{-15}*190}{\left(1.5.10^{-5}\right)^2}\frac{1}{293} = 2.83.10^{-5}$$

$$Gr\Pr = 2.83.10^{-5} = 1.98.10^{-5}$$

We are in laminar flow!

$$Nu_D = 0.53\left(Gr\,\mathrm{Pr}\right)^{0.25}$$

The resulting flow density is:

$$\varphi_W = \frac{\lambda\, Nu_D}{D}\left(T_W - T_e \right) = \frac{0.0275*0.53*\left(1.98.10^{-5}\right)^{0.25}}{10^{-5}}\,190$$

$$\varphi_W = 1.847.10^4\ W\,m^{-2}$$

which is to be compared to $\varphi_W = 4318*190 = 8.34.10^5\ W\,m^{-2}$ in forced convection.

2) Impulse thermal anemometer

The convection coefficient between the wire remains $h = 4318\ W\,m^{-2}\,K^{-1}$.

The instantaneous thermal balance for a length of wire l is expressed by:

$$\rho c \frac{\pi D^2}{4} l\, dT = \pi D\, l\, \varphi_W$$

Or also:

$$\rho c \frac{\pi D^2}{4} l\, dT_W = -\pi D\, l\, h\left[T_W(t) - T_e \right]$$

The temperature $T_W(t)$ is assumed to be homogeneous throughout the wire, which is highly conductive.

This gives a differential equation:

$$\frac{d\left[T_W(t) - T_e \right]}{dt} = -\alpha \left[T_W(t) - T_e \right]$$

with: $\alpha = \dfrac{4h}{\rho c D}$

We used the fact that T_e is constant, so $\dfrac{dT_e}{dt} = 0$.

Let us solve this equation using a traditional method:

$$\frac{1}{T_W(t) - T_e} \frac{d\left[T_W(t) - T_e\right]}{dt} = \alpha$$

with the boundary condition:

$$t = 0; \quad T_W - T_e = T_{W0} - T_e$$

$$Ln\left[T_W(t) - T_e\right] = -\alpha t + Ln\,C$$

$$T_W(t) - T_e = C e^{-\alpha t}$$

$$T_W(t) - T_e = \left(T_{W0} - T_e\right) e^{-\alpha t}$$

The time constant of this exponential is therefore:

$$\tau = \frac{1}{\alpha} = \frac{\rho c D}{4h}$$

The value of τ emerges:

$$\tau = \frac{21400 * 140 * 10^{-5}}{4 * 4318} = 1.73.10^{-3} = 1.73\,ms$$

Even though we measure the half-period, we cannot expect better than $f = \frac{2}{\tau} = 1{,}150\ Hz$.

NOTE.– The spectral study of signals is central to the study of turbulence. Sensors with very short response time are required. In practice, this value in the range of $f = 1{,}000\ Hz$ is unsatisfactory. In the turbulent spectral analysis, we seek to reach frequencies higher than $f = 10{,}000\ Hz$.

The hot wire system is much more commonly used than the thermal anemometer.

To reduce the thermal inertia of the wire, we will try to use even finer wires (in the micrometer range), which are very fragile and delicate to use.

A first method consisted historically of measuring the temperature of the wire at constant intensity. The cut-off frequency of the measurement was then determined by the thermal inertia of the wire. A more efficient technique was soon used by feeding the wire with a circuit that adjusts the intensity of the wire, in order to maintain its constant temperature. The thermal inertia of the wire has thus been replaced by the much less annoying thermal inertia of an electronic circuit.

These considerations also explain why we turned to the optical measurement of velocities (Laser-Doppler anemometry) a few decades ago.

EXAMPLE 2.3.– Forced convection: orders of magnitude

1) We consider a tube with an internal diameter of $D = 5\text{cm}$. Its temperature is $T_W = 80°\text{C}$.

We make it go through four flows:

- two air flows at $T_e = 20°\text{C}$ with respective flow rates;

$q_{V1} = 1\,m^3\,hr^{-1}$

and $q_{V2} = 2200\,m^3\,hr^{-1}$

- two water flows at $T_e = 20°\text{C}$ with respective flow rates;

$q_{V3} = 70\,lit\,hr^{-1}$

and $q_{V4} = 144\,m^3\,hr^{-1}$.

For each flow, calculate the parietal heat flow density exchanged between the wall and the fluid.

2) We consider two plane parallel plates separated by the distance $e = 5\,cm$ and whose temperature is still $T_W = 80°\text{C}$.

Four rectilinear flows, two of air and two of water at $T_e = 20°C$, which have the same respective flow velocities as the four flows in question 1, are made to traverse the space between these two plates.

For these four flows, calculate the parietal heat flow density.

SOLUTION TO EXAMPLE 2.3.–

1) Let us first determine the regimes of these four flows.

The section of the tube is: $S = \frac{\pi D^2}{4} = 1.96.10^{-3}\ m^2$.

The four flowing speeds are as follows:

$$V_{q1} = \frac{q_{V1}}{S} = \frac{1}{3600*1.96.10^{-3}} = 0.142\, m\, s^{-1}$$

$$V_{q2} = \frac{q_{V2}}{S} = \frac{2200}{3600*1.96.10^{-3}} = 311.8\, m\, s^{-1}$$

$$V_{q3} = \frac{q_{V3}}{S} = \frac{70.10^{-3}}{3600*1.96.10^{-3}} = 9.9.10^{-3}\ m\, s^{-1}$$

$$V_{q4} = \frac{q_{V4}}{S} = \frac{144}{3600*1.96.10^{-3}} = 20.4\, m\, s^{-1}$$

The corresponding Reynolds numbers will be:

$$R_{D1} = \frac{V_{q1} D}{\nu} = \frac{0.142*0.05}{1.5.10^{-5}} = 473$$

The flow regime is laminar.

$$R_{D2} = \frac{V_{q2} D}{\nu} = \frac{311.8*0.05}{1.5.10^{-5}} = 1.04.10^{6}$$

The flow regime is turbulent.

$$R_{D3} = \frac{V_{q1} D}{\nu} = \frac{9.9.10^{-3} * 0.05}{1.10^{-6}} = 495$$

The flow regime is laminar.

$$R_{D4} = \frac{V_{q4} D}{\nu} = \frac{20.4 * 0.05}{1.10^{-6}} = 1.02.10^{6}$$

The flow regime is turbulent.

Since we are looking for orders of magnitude, we can take the properties of the fluids at their core temperature, that is, $T_e = 20\,°\text{C}$.

Let us calculate the convection coefficients in each of the four cases, then the parietal flow densities:

For laminar regimes, we will have:

$$h = \frac{\lambda\, Nu_D}{D} = \frac{3.66\,\lambda}{5.10^{-2}} = 73.2\,\lambda$$

$$\varphi_W = h\left(T_W - T_e\right) = 60 * h$$

For turbulent regimes, we will have:

$$h = \frac{\lambda\, Nu_D}{D} = \frac{\lambda\left(0.023\, R_D^{0.8}\, P_r^{1/3}\right)}{D} = \frac{0.023\,\lambda}{D^{0.2}} \left(\frac{V_q}{\nu}\right)^{0.8} P_r^{1/3}$$

$$h = \frac{\lambda\, Nu_D}{D} = 4.19.10^{-2}\,\lambda \left(\frac{V_q}{\nu}\right)^{0.8} P_r^{1/3}$$

$$\varphi_W = h\left(T_W - T_e\right) = 60 * h$$

Let us apply this to our four cases:

Air flow $q_{V1} = 1\, m^3\, hr^{-1}$

$$h = 73.2\,\lambda = 73.2 * 0.0257 = 1.88\ W\,m^{-2}\,K^{-1}$$

$$\varphi_{W1} = h_1\left(T_W - T_e\right) = 60 * h_1 = 113\ W\,m^{-2}$$

Air flow $q_{V2} = 2200\ m^3\,hr^{-1}$

$$h_2 = 4.19.10^{-2} * 0.0257\left(\frac{311.8}{1.5.10^{-5}}\right)^{0.8} 0.7^{1/3} = 683\,W\,m^{-2}\,K^{-1}$$

$$\varphi_W = h_2\left(T_W - T_e\right) = 60 * 683 = 4.1.10^4\ W\,m^{-2}$$

Water flows $q_{V3} = 70\,lit\,hr^{-1}$

$$h_3 = \frac{\lambda\,Nu_D}{D} = 73.2 * 0.597 = 43.7\ W\,m^{-2}\,K^{-1}$$

$$\varphi_{W3} = h_3\left(T_W - T_e\right) = 60 * 43.7 = 2622\ W\,m^{-2}$$

Water flows $q_{V4} = 144\ m^3\,hr^{-1}$

$$h_4 = 4.19.10^{-2} * 0.597\left(\frac{20.4}{10^{-6}}\right)^{0.8} 7^{1/3} = 3.37.10^4\ W\,m^{-2}\,K^{-1}$$

$$\varphi_{W4} = h_4\left(T_W - T_e\right) = 60 * 3.37.10^4 = 2.02.10^6\ W\,m^{-2}$$

2) Flow between two plates

If we compare the formulas giving the Nusselt for each geometry and each flow regime, we must also compare the following in a laminar regime:

For the pipe at $Nu_D = 3.66$, for the plates $Nu_e = 3.4$, in other words, a ratio of $\frac{3.4}{3.66} = 0.929$.

In a turbulent regime:

For the pipe at $Nu_e = 0.023\ R_e^{0.8}\ \mathrm{Pr}^{1/3}$, for the plates $Nu_e = 0.023\ R_e^{0.8}\ \mathrm{Pr}^{1/3}$, that is, the same Colburn formula.

In the case of plates, the reference length is the distance e of the plates. For this example, e is equal to the diameter D of the tube. Only the initial coefficients modify the results.

The results for the plates will therefore be deduced very simply:

Air flow $q_{V1} = 1\, m^3\, hr^{-1}$

$$\varphi_{W1} = 113 * 0.929 = 105\, W\, m^{-2}$$

Air flow $q_{V2} = 2200\, m^3\, hr^{-1}$

$$\varphi_{W2} = 4.1.10^4\, W\, m^{-2}$$

Water flows $q_{V3} = 70\, lit\, hr^{-1}$

$$h = \frac{\lambda\, Nu_D}{D} = 73.2 * 0.597 = 43.7\, W\, m^{-2}\, K^{-1}$$

$$\varphi_{W3} = 2622 * 0.929 = 2436\, W\, m^{-2}$$

Water flows $q_{V4} = 144\, m^3\, hr^{-1}$

$$\varphi_{W4} = 2.02.10^6\, W\, m^{-2}$$

EXAMPLE 2.4.– Thermal balance of a roof under the wind

A roof slope (see Figure 2.1) has a dimension of $L_1 = 16\, m$ in the horizontal direction and $L_2 = 7\, m$ in the vertical direction. It is swept by a small breeze, in other words, a wind speed of $U = 19\, km.hr^{-1}$, normal to its horizontal dimension. The roof is covered with slates. It will be assimilated to a gray body of emissivity $\varepsilon = 0.9$. The solar radiation at the time of day considered is $\varphi_S = 300\, W\, m^2$.

We recall the Stefan constant: $\sigma = 5.67.10^{-8}\, SI$.

We think in terms of average convection coefficient here.

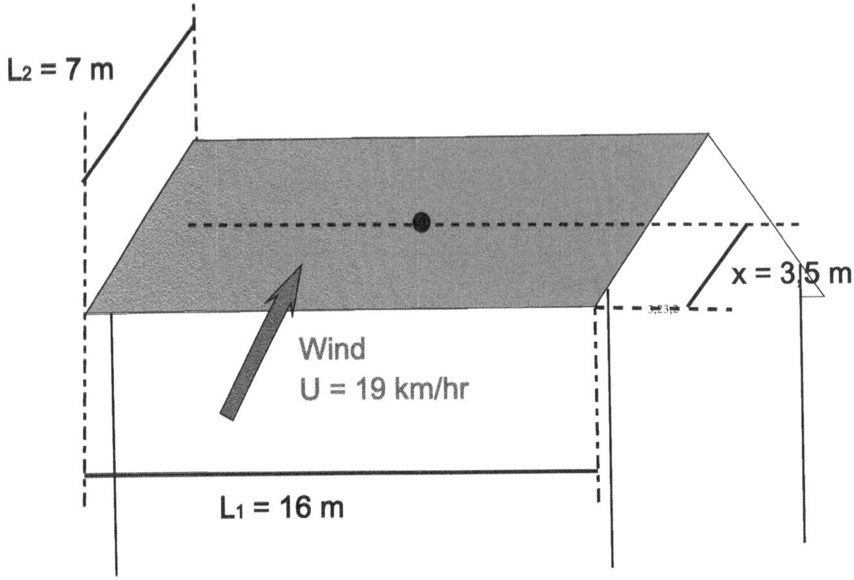

Figure 2.1. *Heat transfers on a roof. For a color version of this figure, see www.iste.co.uk/ledoux/heat3.zip*

1) The owner of the house finds the slates very hot; they first assume that the temperature of the roof is consistent and equal to $T_{W1} = 40°C$. The ambient air has a temperature of $T_a = 20°C$, of which, for simplicity, we will take the physical properties of the air.

1.1) Calculate the thermal flow carried by the wind on the considered roof slope.

1.2) What is the flow radiated by the roof? What is the absorbed flow?

1.3) What do you think about the temperature $T_{W1} = 40°C$ here?

2) We want to assess the real value of the temperature of the roof, T_{W2}, which we will consider to be consistent. Considering a balance between radiation and convection, what is the value of T_{W2}?

3) The previous calculation assumes a consistent roof temperature. We know that in a boundary layer, the convection coefficient is not consistent anyway. We place ourselves at a distance $x = 1\,m$ from the lower edge of the roof. Assess the local temperature T_{W3} under the assumptions of question 2.

What should we make of these different temperature assessments?

4) The impact of natural convection has been neglected here. Should the assessments made here be challenged?

5) Conclusion?

SOLUTION TO EXAMPLE 2.4.–

1) Calculation of convection at an assumed parietal temperature $T_{W1} = 40°C$.

1.1) Let us establish the table of properties of air to be used:

The values of the physical properties of the air will be taken at $T_a = 20°C$.

Density $\rho = 1.2\ kg.m^{-3}$

Dynamic viscosity $\mu = 1.81.10^{-5}\ Pl$

Thermal conductivity $\lambda = 0.0257\ Wm^{-1}K^{-1}$

Mass heat capacity at constant pressure $c_P = 1006\ J\ kg^{-1}$

From which we can deduce:

Kinematic viscosity $\nu = \dfrac{\mu}{\rho} = \dfrac{1.81.10^{-5}}{1.2} = 1.5.10^{-5}\ m^2\ s^{-1}$

Thermal diffusivity $a = \dfrac{\lambda}{\rho c_P} = \dfrac{0.0257}{1.2*1006} = 2.13.10^{-5}\ m^2\ s^{-1}$

Prandtl number $\Pr = \dfrac{\nu}{a} = \dfrac{1.5.10^{-5}}{2.13.10^{-5}} = 0.7$

We need to determine the flow regime.

The Reynolds at the roof peak will be equal to:

$$U = 19\ km.hr^{-1} = \frac{19000}{3600} = 5.28\ m\ s^{-1}$$

$$R_L = \frac{U\,L}{\nu} = \frac{5.28*7}{1.55.10^{-5}} = 2.38.10^6$$

We will consider that we are in a turbulent regime.

The Nusselt number will be:

$$Nu_L = 0.036\ R_L^{0.8}\ P_r^{1/3}$$

From this, we deduce the average convection coefficient:

$$h = \frac{\lambda}{L} 0.036 \left(\frac{U\,L}{\nu}\right)^{0.8} \Pr^{1/3}$$

$$h = \frac{0.0257}{16} 0.036 \left(\frac{5.28*16}{1.55.10^{-5}}\right)^{0.8} \sqrt[3]{0.72} = \frac{0.0257}{16} 0.036 * 2.45.10^5 * 0.896$$

$$h = 12.69\,W\ m^{-1}\ K^{-1}$$

And the average flow density will be:

$$\varphi_W = h\left(T_{W1} - T_a\right)$$

$$\varphi_W = 12.69\left(40 - 20\right) = 253\ W\ m^{-2}$$

The flow carried by the wind will then be:

$$S = L_1\ L_2 = 16*7 = 112\ m^2$$

$$\Phi = S\,h\left(T_{W1} - T_a\right)$$

$$\Phi = 253*112 = 28.432\,kW$$

1.2) The radiation from the roof will be, per unit area, that of a gray body at absolute temperature:

$$T_{W1} = 40°\text{C} = 313K$$

The radiation flow density is:

$$\varphi_R = \varepsilon \sigma T_{W1}^4 = 0.9 * 5.67.10^{-8} * 313^4 = 490 W\, m^{-2}$$

The flow density received is: $\varphi_S = 500\ W\, m^{-2}$. The density absorbed by the gray body will be:

$$\varphi_{abs} = \varepsilon\, \varphi_S = 0.9 * 500 \ = 450 W\, m^{-2}$$

1.3) The thermal balance at the surface will be, bearing in mind that the absorbed solar heat is divided between roof radiation and wind convection:

$$\varphi_{abs} = \varphi_R + h\left(T_{W1} - T_a\right)$$

We see here that the radiated energy φ_R is greater than the energy brought by the sun φ_{abs}, while the energy carried by the wind $h\left(T_{W1} - T_a\right)$ is obviously positive (the roof cannot have a temperature lower than the environment!).

This result is non-sensical; the temperature has therefore been overestimated, since the calculated radiated energy is obviously too strong.

2) To determine an "average" temperature (uniform over the entire roof surface), let us rewrite the issue from question **1.3**, with an unknown temperature T_{W2}:

$$\varphi_{abs} = \varphi_R + h\left(T_{W1} - T_a\right)$$

Or:

$$\varepsilon\, \varphi_S = \varepsilon \sigma T_{W2}^4 + h\left(T_{W2} - T_a\right)$$

T_{W2} is then the solution of a fourth degree algebraic equation that we can solve with a solver:

$$\varepsilon \sigma T_{W1}^4 + h T_{W2} = h T_a + \varepsilon\, \varphi_S$$

$$5.13.10^{-8}\, T_{W1}^4 + 12.69\, T_{W2} = 12.69 * 293 + 450$$

$$5.13.10^{-8}\, T_{W1}^4 + 12.69 T_{W2} - 4168 = 0$$

$$T_{W2} = 296.99\, K = 24°\text{C}$$

3) Local calculation

Let us calculate the local convection coefficient in laminar regime.

The Nusselt number will be:

$$Nu_L \ = 0.0288\, R_L^{0.8}\, P_r^{1/3}$$

We deduce the local convection coefficient:

$$h = \frac{\lambda}{x} 0.0288 \left(\frac{U\,x}{\nu} \right)^{0.8} \text{Pr}^{1/3}$$

where x is the distance to the base of the roof.

For $x = 1\, m$ we find:

$$h_{loc} = \frac{0.0257}{1} 0.0288 \sqrt{\frac{5.28 * 1}{1.55.10^{-5}}}\; 0.72^{1/3} = 17.67\, W\, m^{-2}\, K^{-1}$$

The local thermal balance then gives us:

$$\varepsilon\, \sigma\, T_{W3}^4 + h_{loc}\, T_{W3} = h T_a + \varepsilon\, \varphi_S$$

$$5.13.10^{-8}\, T_{W1}^4 + 17.67\, T_{W2} = 17.67 * 293 + 450$$

$$5.13.10^{-8}\, T_{W1}^4 + 0.446\, T_{W2} - 5627 = 0$$

This results in a local temperature T_{W3} at $x = 1\, m$ to the base of the roof:

$$T_{W3} = 296\, K = 23°\text{C}$$

4) We can also assess the losses by natural convection, which we have neglected here.

Taking the simple example of a horizontal surface, we can find an average Nusselt for the roof. Let us maximize the temperature of the roof to $T_W = 40\,°\mathrm{C}$.

Taking the width as reference length, the Grashof can be assessed at:

$$G_{r1} = \frac{g\, l^3\, \Delta T}{\nu^2\, T_W} = \frac{9.81*(7)^3*(40-20)}{\left(1.5.10^{-5}\right)^2*313} = 9.55.10^{11}$$

We are in a turbulent regime.

We deduce the Nusselt and the average convection coefficient:

$$Nu_{L1} = 0.14\,(Gr\,\mathrm{Pr})^{1/3} = 0.14\left(9.55.10^{11}*0.7\right)^{1/3} = 1224$$

$$h_1 = \frac{\lambda\, Nu_{L1}}{L} = \frac{0.0257*1224}{16} = 1.96\; W\, m^{-2}\, K^{-1}$$

which can be compared to $h_{loc} = 17.67\; W\, m^{-2}\, K^{-1}$ and $h = 12.69 W\, m^{-1}\, K^{-1}$.

The approximation that consists of neglecting the natural convection remains roughly acceptable.

5) Conclusion. We see that the average assessment is sufficient from a practical perspective.

We may want to assess the influence of convection. To do this, let us assume that there is no convection. The thermal balance will then be resolved to the equality of the absorbed and emitted flows. We can deduce a temperature T_{W4}, such as:

$$\varphi_{abs} = \varphi_R = \varepsilon\;\sigma T_{W4}^4$$

Or:

$$T_{W4}^4 = \frac{\varphi_R}{\varepsilon\;\sigma} = \frac{450}{0.9*5.67.10^{-8}} = 8.818.10^9$$

$$T_{W4} = \left(8.818.10^9\right)^{0.25} = 306\ K = 33.4°C$$

We can see that convection has a significant influence on the thermal balance and the determination of temperature. This could have been predicted by comparing the flow densities found above.

Furthermore, the house owner's intuition turns out to be very doubtful.

EXAMPLE 2.5.– A teapot

A teapot (see Figure 2.2) is similar to a metal cylinder of diameter $D = 10cm$ and height $H = 12cm$. Its lid is a flat disc with a diameter of $D = 10cm$. We consider that the thickness of the metal and its thermal conductivity are such that the temperature gradient between the inner and outer faces of the teapot is negligible.

The tea is prepared by filling the teapot with water at a temperature of $T_i = 98°C$. The room temperature is $T_e = 21°C$. The ambient air is assumed to be calm.

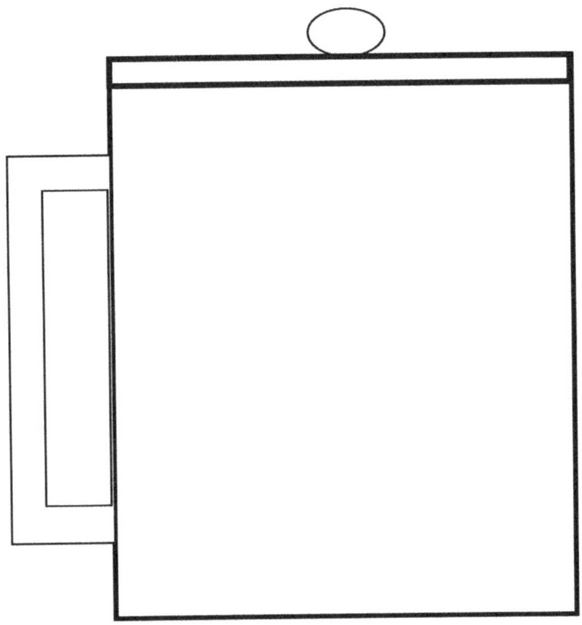

Figure 2.2. *Diagram of the teapot*

1) Calculate the initial heat loss flow of the teapot, taking into account the side surface and the lid. We will neglect any transfer to the table, as well as the handle.

For the sake of simplicity, we will assume that the side wall of the teapot is a simply curved plane.

2) How many degrees does the temperature of the tea drop when the water is at $T_{i1} = 98\,°C$?

3) How many degrees does the temperature of the tea drop when the water is at $T_{i2} = 40\,°C$?

4) Can you roughly estimate the time it takes for the temperature of the tea to drop from $T_{i1} = 98\,°C$ to $T_{i2} = 40\,°C$?

SOLUTION TO EXAMPLE 2.5.–

1) The heat losses will obviously result from a transfer by natural convection. For the natural wall, there will be convection on a vertical wall and for the lid, convection on a horizontal wall.

Let us **collect** the a priori interesting **numerical values:**

The surfaces to be considered for the transfer are:

For the vertical wall: $S = \pi D H = \pi * 10^{-1} * 0.12 = 3.77.10^{-2}\ m^2$.

For the lid: $S = \pi \dfrac{D^2}{4} = \dfrac{\pi * 10^{-2}}{4} = 7.85.10^{-3}\ m^2$.

Here are the important physical properties:

For the air:

$$\nu = \frac{\mu}{\rho} = \frac{1.81.10^{-5}}{1.2} = 1.5.10^{-5}\ m^2 s^{-1}$$

$$\lambda = 0.0257\ W m^{-1} K^{-1}$$

$$\Pr = 0.7.$$

For the water:

Mass heat capacity at constant pressure $c_P = 4182\ J\, kg^{-1}$.

The flows will be of the following form, for each element:

$$\Phi = S\ \varphi_w = S\ \varphi_w\ h\left(T_W - T_e\right) = S\ \varphi_w\ h\left(T_i - T_e\right)$$

We know that, in the cases calculated, the convection coefficient h will be from a Nusselt number using the following relations:

$$Nu_L = C\left(Gr\,\Pr\right)^n$$

$$Nu_L = \frac{h\ L}{\lambda}$$

Laminar regime $n = \dfrac{1}{4}$

Turbulent regime $n = \dfrac{1}{3}$

The form factor C will depend on the product $Gr\,\Pr$.

Let us recall that the Grashof number is written as:

$$G_r = \frac{g\ \beta\ L^3\ \Delta T}{\nu^2}$$

We will take $\beta = \dfrac{1}{T}$, as justified above.

For the vertical wall, we will take the height H as reference length and for the cover, we will take the valid formula for a square with a side equal to the diameter D.

Therefore:

For the vertical wall: $G_{r1} = \dfrac{g\ H^3\ \Delta T}{\nu^2\ T_i}$.

For the lid: $G_{r2} = \dfrac{g\, D^3\, \Delta T}{\nu^2\, T_i}$.

We use T_i as temperature for the assessment of β. This temperature is, indeed, more characteristic of the interior of the boundary layer than the ambient temperature.

Initially, the water is at $T_i = 98\,°\text{C} = 371\,K$.

Considering the vertical wall as an uncurved wall, we will have the following by consulting our form:

For the vertical wall: $G_{r1} = \dfrac{g\, H^3\, \Delta T}{\nu^2\, T_i} = \dfrac{9.81 * (0.12)^3 * (98-21)}{\left(1.5.10^{-5}\right)^2 * 371} = 1.56.10^7$.

For the lid: $G_{r1} = \dfrac{g\, D^3\, \Delta T}{\nu^2\, T_i} = \dfrac{9.81 * (0.1)^3 * (98-21)}{\left(1.5.10^{-5}\right)^2 * 371} = 9.05.10^6$.

We are therefore in a laminar regime in both cases, respecting the same convention of indices as for the Grashof:

$$Nu_{L1} = 0.59\,(Gr\,\Pr)^{1/2} = 0.59\left(1.56.10^7 * 0.7\right)^{1/2} = 1950$$

$$Nu_{L2} = 0.14\,(Gr\,\Pr)^{1/3} = 0.54\left(9.05.10^6 * 0.7\right)^{1/2} = 1359$$

We can already see that the majority of it will be a lateral transfer.

Hence, the following convection coefficients:

$$h_1 = \frac{\lambda\, Nu_{L1}}{L} = \frac{0.0257 * 1950}{0.12} = 417.6\, W\, m^{-2}\, K^{-1}$$

$$h_2 = \frac{\lambda\, Nu_{L2}}{D} = \frac{0.0257 * 1359}{0.1} = 349\, W\, m^{-2}\, K^{-1}$$

The corresponding flows will be:

$$\Phi_1 = S_1\,\varphi_{w1} = S_1\,h_1\left(T_i - T_e\right) = 3.77.10^{-2} * 417.6 * 77 = 1212\,W$$

$$\Phi_2 = S_2\,\varphi_{w2} = S_2\,\varphi_{w2}\,h_2\left(T_i - T_e\right) = 7.85.10^{-3} * 349 * 77 = 211\,W$$

2) When the water is at $T_i = 98°\text{C} = 371\,K$, let us calculate the cooling during one second:

The volume of water in the full teapot is:

$$V_{OL} = \pi\frac{D^2}{4}H = \frac{\pi * (0.1)^2 * 0{,}12}{4} = 9.42.10^{-4}\,m^3$$

The mass is:

$$m = \rho V_{OL} = 1000 * 9.42.10^{-4} = 0.942\,kg$$

And a simple thermal balance gives us the temperature drop ΔT_S assessed in one second by the total flow Φ_{1+2}:

$$\Phi_{1+2} = \Phi_1 + \Phi_2 = 1212 + 211 = 1423\,W$$

$$\Phi_{1+2} = m\,c\,\Delta T_S$$

$$\Delta T_S = \frac{\Phi_{1+2}}{m\,c} = \frac{1423}{0.942 * 4182} = 0.36°\text{C}$$

3) At $T_i = 40°\text{C}$, we can take the same steps in the calculation of the flows and the balance.

For the vertical wall: $G_{r1} = \dfrac{g\,H^3\,\Delta T}{\nu^2\,T_i} = \dfrac{9.81 * (0.12)^3 * (40 - 21)}{\left(1.5.10^{-5}\right)^2 * 313} = 4.57.10^6$

For the lid: $G_{r1} = \dfrac{g\,D^3\,\Delta T}{\nu^2\,T_i} = \dfrac{9.81 * (0.1)^3 * (40 - 21)}{\left(1.5.10^{-5}\right)^2 * 313} = 2.65.10^6$

We are therefore in a laminar regime in both cases, respecting the same convention of indices as for the Grashof:

$$Nu_{L1} = 0.59\left(Gr\,\mathrm{Pr}\right)^{1/2} = 0.59\left(4.57.10^{6} * 0.7\right)^{1/2} = 1055$$

$$Nu_{L2} = 0{,}14\left(Gr\,\mathrm{Pr}\right)^{1/3} = 0.54\left(2.65.10^{6} * 0.7\right)^{1/2} = 735$$

We can already see that the majority of it will be a lateral transfer.

Hence, the following convection coefficients:

$$h_1 = \frac{\lambda\,Nu_{L1}}{L} = \frac{0.0257 * 1055}{0.12} = 226\,W\,m^{-2}\,K^{-1}$$

$$h_2 = \frac{\lambda\,Nu_{L2}}{D} = \frac{0.0257 * 735}{0.1} = 189\,W\,m^{-2}\,K^{-1}$$

The corresponding flows will be:

$$\Phi_1 = S_1\,\varphi_{w1} = S_1\,h_1\left(T_i - T_e\right) = 3.77.10^{-2} * 226 * 19 = 162\,W$$

$$\Phi_2 = S_2\,\varphi_{w2} = S_2\,\varphi_{w2}\,h_2\left(T_i - T_e\right) = 7.85.10^{-3} * 189 * 19 = 28.2\,W$$

The cooling will occur during one second:

The volume of water in the full teapot is:

$$V_{OL} = \pi\frac{D^2}{4}H = \frac{\pi * (0.1)^2 * 0.12}{4} = 9.42.10^{-4}\,m^3$$

The mass is:

$$m = \rho V_{OL} = 1000 * 9.42.10^{-4} = 0.942\,kg$$

And a simple thermal balance gives us the temperature drop ΔT_S assessed in one second by the total flow Φ_{1+2}.

$$\Phi_{1+2} = \Phi_1 + \Phi_2 = 162 + 28.2 = 190.2\,W$$

$$\Phi_{1+2} = m\,c\,\Delta T_S$$

$$\Delta T_S = \frac{\Phi_{1+2}}{m\,c} = \frac{190.2}{0.942 * 4182} = 0.048\,°\mathrm{C}$$

4) We can regulate the cooling time of the teapot from $T_{i1} = 98\,°\mathrm{C}$ **to** $T_{i2} = 40\,°\mathrm{C}$ by noting that the cooling per second of the water will decrease with the temperature of the water, thus with time.

This time will therefore inevitably be greater than $t_{R1} = \dfrac{40}{0.36} = 111\,s = 1\,mn\;51\,s$ and less than $t_{R2} = \dfrac{40}{0.048} = 833\;s = 13\,mn\,5\,s$.

NOTE.– This method is definitely imprecise.

To obtain an exact result, it would be necessary to establish and solve the differential equation, which the interior temperature obeys. It should then be noted that the exponential solution, most often obtained in quasi-stationary calculations for the temperature differential (see the parent work devoted to conduction, Chapter 4), will not be valid here. Indeed, the Nusselt is a function of a Grashof, which is itself proportional to the temperature differential, and also inversely proportional to the absolute temperature of water.

We can no longer write a differential equation that would be answered by the temperature differential alone. The analytical determination of the water temperature becomes particularly complex; we will therefore give up here.

We can infer that the cooling time will be closer to t_{R2} than t_{R1}. This will allow for some chit-chat while drinking the tea, which will hardly be drinkable at temperatures of around $50\,°\mathrm{C}$!

3

The Boundary Layer

3.1. Introduction

This chapter is primarily aimed at external convection, and more directly at convection on a plane wall. However, these developments are an essential prerequisite for the study of transfers on curved walls. Moreover, the general concepts can, to some extent, be applied to the problems of interior transfers.

From a theoretical point of view, this chapter is relatively more challenging for the reader than other parts of this book. However, in order to make the reading more direct for those who do not have a complete understanding of fluid mechanics and its general equations, we have moved the details of the calculations to Appendix 4 and Appendix 5. The study of these appendices is highly recommended for a thorough understanding of these subjects.

We will address two topics here. The first will involve the direct approach of the boundary layer, by solving the three differential equations derived from the principles of fluid mechanics: continuity, momentum (fundamental principle of dynamics) and energy (first principle of thermodynamics). This will be looked at in section **3.1**.

The second topic will also involve an approximate, but very efficient method, which is sometimes a little too forgotten in textbooks: the integral method, which will be the subject of section **3.2**.

3.2. The notion of a boundary layer

An important class of problems resides in the interaction of a viscous flow with a boundary: open (plate, etc.) or closed (tube, etc.) solid boundary, stationary fluid, other flow or other fluid.

These flows have common characteristics, which group them in the boundary layers category:

– ***external boundary layers***: interaction of a flow of indefinite extension with a wall, plane, curve, cylinder, etc.;

– ***internal boundary layers***: interaction of a flow of finite extension with a partially or totally closed wall: pipe flow, channel flow, etc.

– ***"mixed" boundary layers***: jets, etc.

These flows can be laminar or turbulent. Viscosity plays a major role.

3.2.1. *Boundary layer characteristics*

Boundary layers are essentially "thin" flows. Consequently, in any boundary layer, a distinction is made between "lateral" and "longitudinal" scales. More precisely, the parameters (velocities, temperatures, concentrations, etc.) will have much stronger gradients along the lateral dimensions than along the longitudinal dimensions.

For the outer boundary layers, a two-step calculation scheme is encountered:

a) When a solid body is placed in a uniform flow, ***a so-called "potential flow" is calculated*** (see kinematics reminders) from a perfect fluid theory.

b) The viscosity in the ***"thin" boundary layer*** is then taken into account.

The "interior" boundary layers reside in the regime establishment zones. The scales to be compared are then the radius R of a pipe versus the length L of this re-entry zone.

The jet boundary layers compare the lateral dimension of a jet, materialized by a lateral velocity profile, and the longitudinal dimension of the jet.

In this manual, the focus for analytical approaches will be on the external or internal boundary layers. The numerical approach is more appropriate for jet problems.

A prototype of the external boundary layers lies in the flow developed by a flat plate in a uniform velocity flow U_e [or locally uniform if $U_e = U_e(x)$ varies axially]. This is often referred to as the "plate in the path of the wind". In this problem, a distinction is made between an undisturbed flow (or "potential" flow; this problem has been previously discussed) and a boundary layer flow, where the connection between the flow and the parietal zone takes place. This thickness connection zone δ has the property of being "thin" with respect to the longitudinal dimension L of the body considered.

Thus, longitudinal scales (according to Ox, in Cartesian coordinates) are always larger than lateral scales (according to Oy, in Cartesian coordinates).

In orders of magnitude:

$$\delta << L \quad [3.1]$$

Two consequences result in the *orders of magnitude* of the terms in the fluid mechanics equations:

The boundary layer is "thin", so the flow is "almost parallel to the flow".

In Cartesian coordinates and in plane flow, we have a first relation between the longitudinal u and lateral v components of the velocity: $u << v$.

The longitudinal and lateral derivatives of the same function f will also have the same ratio of orders of magnitude.

$$\frac{\partial f}{\partial x} << \frac{\partial f}{\partial y} \quad [3.2]$$

The fluid mechanics equations will be simplified, but within limits.

Indeed, in orders of magnitude:

$$\frac{\partial u}{\partial x} \approx \frac{u}{L} \quad ; \quad \frac{\partial v}{\partial y} \approx \frac{v}{\delta} \quad ; \quad v << u \quad ; \quad \delta << L \quad [3.3]$$

$$\text{but:} \quad \frac{u}{L} \approx \frac{v}{\delta} \quad ; \quad \frac{\partial u}{\partial x} \approx \frac{\partial v}{\partial y} \quad [3.4]$$

$$u\frac{\partial u}{\partial x} \approx u\frac{u}{L} \quad ; \quad v\frac{\partial u}{\partial y} \approx u\frac{v}{\delta} \quad ; \quad \frac{\partial^2 u}{\partial y^2} \approx \frac{u}{\delta^2} \quad [3.5]$$

These three terms are of the same order of magnitude.

The establishment of the boundary layer equations is detailed in **Appendix 5**.

For the most part in the following, we will use a simplified version of these equations, responding to assumptions that are often met:

– the fluid is incompressible;

– the properties of the fluid are invariant, especially with temperature;

– the flow is not fast enough to lead to kinetic heating.

We will thus have an incompressible and stationary plane flow over a flat plate:

$$\frac{\partial u}{\partial x}+\frac{\partial v}{\partial y}=0 \quad [3.6]$$

$$u\frac{\partial u}{\partial x}+v\frac{\partial u}{\partial y}=\nu\frac{\partial^2 u}{\partial y^2}$$

$$u\frac{\partial T}{\partial x}+v\frac{\partial T}{\partial y}=\nu\frac{\partial^2 T}{\partial y^2} \quad [3.7]$$

Note that not all derivation terms in x disappear, because these equations contain products of small terms $\left(\frac{\partial u}{\partial x},\frac{\partial T}{\partial x},v\right)$ by large terms $\left(u,\frac{\partial u}{\partial y},\frac{\partial^2 u}{\partial y^2},\frac{\partial T}{\partial y},\frac{\partial^2 T}{\partial y^2}\right)$.

As a physicist, it is important to approach these notions of order of magnitude.

In the external boundary layers, once we move away from the wall, we quickly reach a speed close to that of the potential flow, U_e. Strictly speaking, the variation is asymptotic, so we will say that u tends to U_e, when x tends to infinity. In wind tunnels, we will thus have infinities a few centimeters long!

Boundary layer thickness δ does not mean infinitely small.

In a wind tunnel, for a plate that is a few dozen centimeters long, the δ scale is the centimeter.

If we take a ship of length $L = 250\,m$, the δ scale will be the meter.

At the scale of a territory, the atmospheric boundary layer can measure one to two kilometers!

Moreover, the infinitesimally small δ before L is often reduced to ratios $\frac{\delta}{L}$ in the range of 10^{-2}.

3.2.2. *The boundary layers can be approached by different methods*

For laminar flows, boundary layer equations can be solved. Several examples are given below: Blasius theory, and Pohlhausen theory for the thermal problem.

For turbulent flows, different approaches exist: semi-empirical theories, mixing length theory and Boussinesq viscosity, numerical approaches (Reynolds decomposition, $k-\varepsilon$ methods, etc.) We refer to the literature, which is very rich on the subject.

We can also obtain reasonable results at a "lower cost" by using integral methods. This approach, which is quite old, is still interesting to know. We wanted to introduce the interested reader to these techniques and devote a specific section to it.

3.3. The external boundary layers: analytical treatment

3.3.1. *The laminar boundary layer developed by a flat plate in a uniform flow*

This problem is still referred to as the flat plate "in the path of the wind". *Although laminar flows may seem rather academic from a practical point of view, the reasoning developed on this model is very stimulating and formative for the thought process behind fluid mechanics.*

We consider a fluid with constant physical properties (density μ and kinematic viscosity v). The potential flow is then obviously a uniform flow with speed U_e. We place ourselves in the case of a stationary plane flow.

The system (Figure 3.1) is reduced to an orthonormal reference frame. The origin is at the leading edge of the plate, the x axis is parallel and the y axis is the normal to this plate.

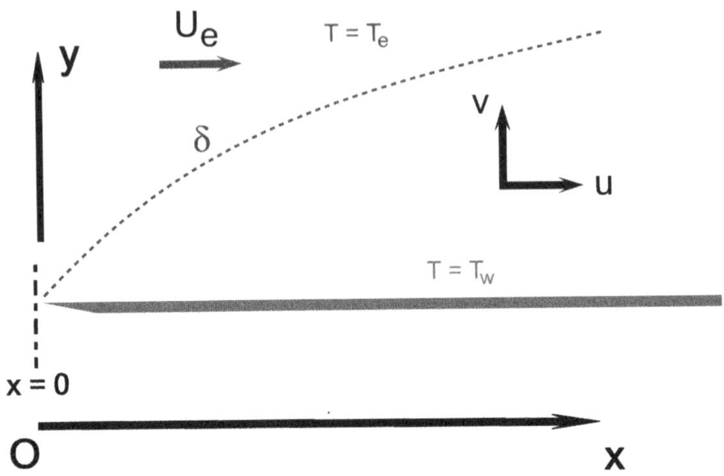

Figure 3.1. *Flat plate in the path of the wind. For a color version of this figure, see www.iste.co.uk/ledoux/heat3.zip*

We have seen that the velocity vector $\vec{V}(u,v)$ answers the following equations.

Continuity equation

$$\frac{\partial u}{\partial x}+\frac{\partial v}{\partial y}=0 \qquad [3.8]$$

Momentum equation

$$u\frac{\partial u}{\partial x}+v\frac{\partial v}{\partial y}=-\frac{1}{\rho}\frac{d\,p_G}{d\,x}+\nu\frac{\partial^2 u}{\partial y^2} \qquad [3.9]$$

Energy equation

$$u\frac{\partial T}{\partial x}+v\frac{\partial T}{\partial y}=\nu\frac{\partial^2 T}{\partial y^2} \qquad [3.10]$$

3.3.1.1. *Blasius problem*

Blasius solved the problem of the laminar boundary layer of a fluid, with constant physical properties over a flat plate of indefinite width, placed in the bed of a wind consisting of a uniform flow. Pohlhausen added a transformation of the energy equation.

The system of the three equations, continuity, momentum and energy, with three unknowns $u\ (x, y)$, $v\ (x, y)$ and $T\ (x, y)$ becomes:

$$\frac{\partial u}{\partial x}+\frac{\partial v}{\partial y}=0 \qquad [3.11]$$

$$u\frac{\partial u}{\partial x}+v\frac{\partial v}{\partial y} = -\frac{1}{\rho}\frac{d\,p_e}{d\,x}+\nu\frac{\partial^2 u}{\partial y^2} \qquad [3.12]$$

$$u\frac{\partial T}{\partial x}+v\frac{\partial T}{\partial y} = a\frac{\partial^2 T}{\partial y^2} \qquad [3.13]$$

The term $-\frac{1}{\rho}\frac{d\,p_e}{d\,x}$ calls for special attention. It comes from the term $-\vec{g}rad\ p$ in the general equations. A boundary layer is a flow that is very close to a parallel flow. Perpendicular to the wall, we can therefore assume a zero gradient of generating pressure $p_G = p + \rho g Z$, where Z is the coordinate of a point on a vertical axis directed from bottom to top.

We therefore have $\frac{\partial(p+\rho g Z)}{\partial y}=0$. [3.14]

For a non-weighted fluid (gas), the front term is negligible and we have:

The pressure is thus constant for a datum. It will therefore be equal to the pressure in the external flow.

Moreover, in this external flow, the fluid is supposed to be perfect. Bernoulli's theorem therefore applies. If the speed of the external flow $U_e\ (x)$ is variable according to x, we can write:

$$d\left(\frac{\rho U_e^2}{2}+p_e\right)=0 \qquad [3.15a]$$

This allows us to rewrite the pressure term as:

$$-\frac{1}{\rho}\frac{d\,p_e}{d\,x}=U_e\frac{d\,U_e}{d\,x} \qquad [3.15b]$$

Thus, we often find the momentum equation in the following form:

$$u\frac{\partial u}{\partial x}+v\frac{\partial v}{\partial y} = U_e\frac{d\,U_e}{d\,x}+\nu\frac{\partial^2 u}{\partial y^2} \qquad [3.16]$$

If the external flow is uniform, then the pressure term cancels out and the momentum equation becomes:

$$u\frac{\partial u}{\partial x}+v\frac{\partial v}{\partial y} = \nu\frac{\partial^2 u}{\partial y^2} \qquad [3.17]$$

The boundary conditions are written as:

$$y=0 \quad ; \quad u=v=0 \qquad [3.18]$$

$$y\rightarrow\infty \quad ; \quad u\rightarrow U_e \qquad [3.19]$$

From this point, we can already see the symmetrical roles of ν and a that will now highlight their ratio, which is the Prandtl number:

$$\mathrm{Pr}=\frac{\nu}{a} \qquad [3.20]$$

We will note that the problem poses an asymptotic variation of the velocity towards the external flow "at infinity". In fact, we will see that the convergence is fast. We define a boundary layer thickness by the point where $u\,(x,\delta)=0.99U_e$. This implies $\delta=\delta(x)$. In the following, we will verify that $\frac{\delta(x)}{x}$ is low.

We will deal with the rest in the context of this last hypothesis.

Blasius showed that the problem becomes self-similar by defining a composite dimensionless variable $\eta(x,y)$:

$$\eta=y\sqrt{\frac{U_e}{\nu x}} \qquad [3.21]$$

Indeed, we perform a change of coordinates from the (x, y) system to the (ξ, η) system, with the following definitions:

$$\xi = x \qquad [3.22]$$

$$\eta = y\sqrt{\frac{U_e}{\nu x}} \qquad [3.23]$$

A function $f(\xi,\eta)$ (which will turn out to be a unique function of η) then appears:

$$f(\xi,\eta) = \frac{\psi}{\sqrt{\nu U_e x}} \qquad [3.24]$$

We will also show a reduced temperature $\theta(x,y)$ defined by:

$$\theta(x,y) = \frac{T(x,y) - T_w}{T_e - T_w} \qquad [3.25]$$

By a rather long calculation that we have reported in **Appendix 5**, the so-called Blasius equation emerges:

$$2f''' + ff'' = 0 \qquad [3.26]$$

Assorted boundary conditions:

$$f(0) = 0 \;;\; f'(0) = 0 \;;\; f'(\infty) \to 1 \qquad [3.27]$$

Note here the symmetrical roles of f' and θ.

$f'(\eta)$ is obtained in the form of a series that is calculated numerically.

There are two notable values:

$$f'(4.92) = 0.99 \qquad [3.28]$$

This gives the conventional thickness $\delta(x)$ of the boundary layer:

$$\frac{u}{U_e} = 0.99 \text{ for } y = \delta \qquad [3.29]$$

Therefore, $\eta\left[x, \delta(x)\right] = \frac{\delta}{x}\sqrt{\frac{U_e x}{\nu}} = 4.92$ [3.30]

which we retain in the form:

$$\frac{\delta(x)}{x} = \frac{4.92}{\sqrt{\frac{U_e x}{\nu}}} = 4.92\, R_x^{0.5} \qquad [3.31]$$

We find a boundary layer thickness increasing as $x^{0.5}$.

We also define a Reynolds number based on the boundary layer thickness:

$$R_\delta = \frac{U_e\, \delta(x)}{\nu} \qquad [3.32]$$

Its value is easily seen from the previous calculations:

$$R_\delta = \frac{U_e\, \delta(x)}{\nu} = \frac{U_e\, x}{\nu}\frac{\delta(x)}{x} = R_x * 4.92\, R_x^{-0.5} \qquad [3.33]$$

$$R_\delta = 4.92\, R_x^{0,5} \qquad [3.34]$$

NOTE.– In the early publications on the subject, purists defined the boundary layer thickness as $1^{0/00}$ or $\frac{u}{U_e} = 0.999$. Noting that $f'(4.95) = 0.999$, the conventional boundary layer thickness became:

$$\frac{\delta(x)}{x} = \frac{4.95}{\sqrt{\frac{U_e x}{\nu}}} = 4.95\, R_x^{0.5} \qquad [3.35]$$

This is a formula that we can find. We can see that the asymptotic character of the evolution of the velocity leads to a sufficiently slow convergence, so that the definition "to 1%" is sufficient in practice.

$$f''(0) = 0.33206 \approx 0.332 \qquad [3.36]$$

This allows us to calculate the coefficient of friction:

$$C_f(x) = \frac{2\tau_W}{\rho U_e^2} = \frac{2\mu \left.\dfrac{\partial u}{\partial y}\right|_{y=0}}{\rho U_e^2} = \frac{2\mu \sqrt{\dfrac{U_e}{\nu x}} \dfrac{\partial}{\partial \eta}\left[U_e f'(0)\right]}{\rho U_e^2} \qquad [3.37]$$

So: $$C_f(x) = \frac{2f''(0)}{\sqrt{\dfrac{U_e x}{\nu}}} = 2f''(0) R_x^{-0.5} \qquad [3.38]$$

where $R_x = \dfrac{U_e x}{\nu}$ is the axial Reynolds number.

A formula that we retain in the following form:

$$C_f(x) = \frac{0.664}{\sqrt{R_x}} \qquad [3.39]$$

3.3.1.2. *Polhausen's problem*

Introducing the reduced temperature θ

A calculation shown in **Appendix 5**, leads to the following Polhausen equation:

$$2\theta'' + \Pr f\theta' = 0 \qquad [3.40]$$

Assorted boundary conditions:

$$\theta(0) = 0 \;;\quad \theta(\infty) \to 1 \qquad [3.41]$$

The solution to the Pohlhausen equation is obtained by a relatively simple integration (see Appendix 5):

$$\theta(\eta,\Pr)=\frac{\int_0^{\eta}\left[f''(\bar{\eta})\right]^{\Pr}d\bar{\eta}}{\int_0^{\infty}\left[f''(\bar{\eta})\right]^{\Pr}d\bar{\eta}} \quad [3.42]$$

$\theta(\eta,\Pr)$ can then be calculated numerically.

Determination of the flow

Returning to the definition of $\theta(x,y)=\dfrac{T(x,y)-T_w}{T_e-T_w}$, [3.43]

we can express the parietal flow density:

$$\varphi_W=-\lambda\left.\frac{\partial T}{\partial y}\right|_{y=0}-\lambda\left.\frac{\partial T}{\partial y}\right|_{y=0} \quad [3.44]$$

considering that $\dfrac{\partial}{\partial y}=\sqrt{\dfrac{U_e}{\nu x}}\dfrac{\partial}{\partial\eta}$ [3.45]

$$\varphi_W=-\lambda\left.\frac{\partial T}{\partial y}\right|_{y=0}=-\lambda\left.\frac{\partial(T_e-T_w)\theta}{\partial y}\right|_{y=0}=-\lambda\sqrt{\frac{U_e}{\nu x}}(T_e-T_w)\left.\frac{\partial\theta}{\partial\eta}\right|_{\eta=0} \quad [3.46]$$

Therefore, we must assess:

$$\left.\frac{\partial\theta}{\partial\eta}\right|_{\eta=0}=\frac{\partial}{\partial\eta}\left.\frac{\int_0^{\eta}\left[f''(\bar{\eta})\right]^{\Pr}d\bar{\eta}}{\int_0^{\infty}\left[f''(\bar{\eta})\right]^{\Pr}d\bar{\eta}}\right|_{\eta=0}=\frac{f''^{\Pr}(\bar{0})}{\int_0^{\infty}\left[f''(\bar{\eta})\right]^{\Pr}d\bar{\eta}}=\frac{0.332^{\Pr}(\bar{0})}{\int_0^{\infty}\left[f''(\bar{\eta})\right]^{\Pr}d\bar{\eta}} \quad [3.47]$$

This term can be calculated numerically for different values of $\Pr$.

As can be seen from the following table, the results are very well approximated by:

$$\frac{0.332^{\Pr}(\bar{0})}{\int_0^{\infty}\left[f''(\bar{\eta})\right]^{\Pr}d\bar{\eta}}\approx 0.332\,\Pr^{1/3} \quad [3.48]$$

It is this last expression that the literature will retain.

Pr	**0.6**	**0.7**	**0.8**	**0.9**	**1**	**1.1**	**7**	**10**	**15**
$\dfrac{0.332^{\mathrm{Pr}}(\bar{0})}{\int_0^\infty \left[f''(\bar{\eta})\right]^{\mathrm{Pr}} d\bar{\eta}}$	0.276	0.293	0.307	0.32	0.332	0.334	0.645	0.73	0.835
$0.332\ \mathrm{Pr}^{1/3}$	0.280	0.294	0.308	0.320	0.332	0.342	0.635	0.715	0.818

Table 3.1. *Different values of Pr*

The expression of the flow, the convection coefficient and the Nusselt are immediately deduced:

$$\varphi_W = \lambda\sqrt{\frac{U_e}{\nu x}}(T_e - T_w)0.332\,\mathrm{Pr}^{1/3} \qquad [3.49]$$

$$h = 0.332\,\mathrm{Pr}^{1/3}\,\lambda\sqrt{\frac{U_e}{\nu x}} \qquad [3.50]$$

$$Nu_x = \frac{h\,x}{\lambda} = 0.332\,\mathrm{Pr}^{1/3}\,x\sqrt{\frac{U_e}{\nu x}} \qquad [3.51]$$

So, finally:

$$Nu_x = 0.332\,R_x^{0{,}5}\,\mathrm{Pr}^{1/3} \qquad [3.52]$$

The Stanton number can be deduced from this:

$$St_x = \frac{h}{\rho c_P U_e} = \frac{0.332\,\mathrm{Pr}^{1/3}}{U_e}\,\frac{\lambda}{\rho c_P}\sqrt{\frac{U_e}{\nu x}} \qquad [3.53]$$

Noting that:

$$\frac{\lambda}{\rho c_P} = a = \frac{a\nu}{\nu} = \nu\,\mathrm{Pr}^{-1} \qquad [3.54]$$

$$St_x = \frac{0;332\,\mathrm{Pr}^{1/3}}{U_e} \nu\,\mathrm{Pr}^{-1} \sqrt{\frac{U_e}{\nu\,x}} = 0.332\,\mathrm{Pr}^{1/3}\,\mathrm{Pr}^{-1} \sqrt{\frac{\nu}{U_e\,x}} \quad [3.55]$$

Ultimately:

$$St_x = 0.332\,\mathrm{Pr}^{1/3}\,R_x^{-0.5}\,\mathrm{Pr}^{-2/3} \quad [3.56]$$

This verifies the general relationship established between the Nusselt and the Stanton in **Chapter 2**.

The thickness of the thermal boundary layer δ_{th} is also a point of interest

The ratio of this thickness to the dynamic boundary layer thickness δ is generally sought, also depending on the Prandtl number Pr. A numerical approach shows that the ratio varies depending on the zone of the Prandtl number which we find ourselves in.

For very small Prandtl numbers:

$$\mathrm{Pr} << 1 \quad ; \quad \frac{\delta_{th}}{\delta} \approx \mathrm{Pr}^{-0.5} \quad [3.57]$$

For Prandtl numbers of the order of unity:

$$\mathrm{Pr} \approx 1 \quad ; \quad \frac{\delta_{th}}{\delta} \approx \mathrm{Pr}^{0.38} \quad [3.58]$$

For very large Prandtl numbers:

$$\mathrm{Pr} >> 1 \quad ; \quad \frac{\delta_{th}}{\delta} \approx \mathrm{Pr}^{-0.3} \quad [3.59]$$

This explains why the commonly taken ratio $\frac{\delta_{th}}{\delta}$ is in the range of $\mathrm{Pr}^{-1/3}$. This will notably be the case for air.

Metals have a very large Prandtl number. This leads to very small thermal boundary layer thicknesses, compared to the dynamic boundary layer. The temperature gradients perpendicular to the walls are stronger and the heat flows are higher. This explains why molten metals are very good heat carriers. One of the best known is liquid sodium, which melts at a low temperature. This is appreciated in the nuclear field for this reason (much more from a heat transfer point of view than from a safety point of view; sodium fires are dangerous).

3.3.2. *The turbulent boundary layer*

3.3.2.1. *General approaches*

Of more practical interest, the turbulent boundary layer is not susceptible to the same treatment. The analytical approach can only deal with average parameter values.

a) An important operation is therefore the **calculation of these averages**:

A mean value of velocity is thus defined:

$$\bar{u}\,(x, y) = \frac{1}{T}\int_0^T u\,(t,\, x,\, y) \qquad [3.60]$$

This time average theoretically calculated over an infinite time is, in fact, experimentally established over physically long times (in the range of several minutes).

A so-called Reynolds decomposition follows:

$$u\,(x, y) = \bar{u}\,(x, y) + u' \qquad [3.61]$$

The time average of u' is zero, but not its quadratic average, which leads to evaluating the fluctuations by:

$$\frac{1}{T}\int_0^T u'^2\,(t,\, x,\, y) \qquad [3.62]$$

Also, various correlations will play an important role, such as:

$$\frac{1}{T}\int_0^T u'(t,\, x,\, y)v'(t,\, x,\, y) \qquad [3.63]$$

However, we will see that in the framework of integral methods, similar treatments can be applied to laminar and turbulent layers.

Historically, several approaches have been conducted, generally conditioned by the progress of metrology. Thus, for the boundary layer above a flat plate:

The first studies using the Pitot tube led to the search for average power velocity profiles (which can be used in particular with the integral methods):

$$\frac{\bar{u}(x,y)}{U_e}=\left(\frac{y}{\delta(x)}\right)^n \qquad [3.64]$$

where n is a coefficient that varies in principle as a function of the Reynolds. For the usual Reynolds numbers, n comes out mostly as $\frac{1}{7}$ (law in $\frac{1}{7}$).

With the development of anemometry (hot wire anemometry, then Laser-Doppler anemometry), we sought to render the profiles, and especially the structure of the flow, with more detail.

b) Although this is beyond the scope of this introductory manual, we should note the importance of spectral approaches to fluctuations. This has been made possible by the progressive reduction of sensor response times, from the microscopic hot wire to the optical measurement of velocities (Laser-Doppler anemometry).

We will mention the most significant result here, even though we will not use it later.

We then defined variables x^+, y^+ and dimensionless velocities u^+, v^+ by a reference speed, known as friction velocity:

$$U_f=\sqrt{\frac{2\tau_W}{\rho}}=U_e\sqrt{\frac{C_f}{2}} \qquad [3.65]$$

$$x^+=\frac{U_f\,x}{\nu} \quad ; \quad y^+=\frac{U_f\,y}{\nu} \qquad [3.66]$$

$$u^+=\frac{\bar{u}}{U_f} \quad ; \quad v^+=\frac{\bar{v}}{U_f} \qquad [3.67]$$

The speed profile was then divided into three zones.

The first was the film zone, corresponding to a "viscous" laminar sub-layer with a linear velocity profile.

$$u^+=y^+ \qquad [3.68]$$

This zone has a thickness generally taken at $y^+ = 5$.

The second was the so-called parietal zone, constituting the bulk of the boundary layer. This wall law leads to a logarithmic profile in reduced variables.

$$u^+ = \frac{1}{\kappa} \ln y^+ + C \qquad [3.69]$$

where κ is the von Kármán constant $\kappa \approx 0.4$.

The third was the connection zone to the potential flow.

Using the Reynolds decomposition, the equations for the components of the mean velocity become:

Using the same calculation technique for the instantaneous equation, the equations become the following in their traditional form:

$$\frac{\partial \bar{u}}{\partial x} + \frac{\partial \bar{v}}{\partial y} = 0 \qquad [3.70]$$

$$\bar{u}\frac{\partial \bar{u}}{\partial x} + \bar{v}\frac{\partial \bar{u}}{\partial y} = -\frac{1}{\rho}\frac{\partial \bar{p}_G}{\partial y} + \nu \frac{\partial^2 \bar{u}}{\partial y^2} - \frac{\partial}{\partial y}\left(\overline{u'v'}\right) \qquad [3.71]$$

$$\bar{u}\frac{\partial \bar{T}}{\partial x} + \bar{v}\frac{\partial \bar{T}}{\partial y} = \nu \frac{\partial^2 \bar{T}}{\partial y^2} - \frac{\partial}{\partial y}\left(\overline{T'v'}\right) \qquad [3.72]$$

We cannot therefore reduce the system to these equations of average velocities.

c) We can write an equation for this correlation $\overline{u'v'}$ (we will not do it here). It can then be noted that the process is divergent: for each differential equation written for a correlation of order n, correlations of order $n+1$ appear. This is understandable. The information lost in the statistical approach (creation of terms in u' and v') will not miraculously reappear through simple mathematical manipulations.

It is therefore necessary to close these equations, in other words, to reintroduce information through an additional relation. This closure can be done at different levels of n.

The best known and simplest closure is the gradient closure, as proposed by Boussinesq:

$$\overline{u'v'} = -\varepsilon_V \frac{\partial \bar{u}}{\partial y} \qquad [3.73]$$

where ε_V is defined as a turbulent viscosity. Note that ε_V has the range of a kinematic viscosity.

ε_V is, in general, a function of the velocity field. The literature provides different models that use a different expression of ε_V.

The oldest seems to be the one proposed by Prandtl, who uses a mixing length l_P:

$$\varepsilon_V = l_P \frac{\partial \bar{u}}{\partial y}. \qquad [3.74]$$

Another very common expression uses the kinetic energy of turbulence k and "dissipation" ε (not to be confused with dissipation $\frac{\partial \bar{u}}{\partial y}$ and turbulent kinematic viscosity $\frac{\partial \bar{u}}{\partial y}$) is the basis of the so-called $k-\varepsilon$ model.

It notes that the "dynamic" turbulent viscosity $\mu_T = \rho\varepsilon_V$ is equal to:

$$\mu_T = \rho C_\mu \frac{k^2}{\varepsilon} \qquad [3.75]$$

where C_μ is an empirically determined constant (often $C_\mu \approx 0.09$) that can depend on the problem addressed.

d) The term $\overline{u'v'}$ plays the role of a tangential stress in the equations, and for this reason, it is called the "Reynolds stress". It contains the "turbulent friction" mechanism, illustrated by the "Boussinesq closure".

The viscosity of a fluid is due to the exchange of momentum between layers of fluid in relative translation by molecular agitation. In the same way, the vortices of turbulent flow transport momentum in a velocity gradient. The scale of the phenomenon is, this time, that of the vortex. This is what Prandtl exploited with his theory of "mixing length".

This results in much higher values of turbulent viscosity; the "molecular" viscosity will then often be neglected (except in the laminar sub-layer).

This also means that turbulent viscosity is not a property of the fluid, but of the flow.

e) In the same way, for the energy equation, we will look for a closure in the definition of a turbulent thermal diffusivity ε_θ:

$$\overline{T'v'} = -\varepsilon_\theta \frac{\partial \bar{u}}{\partial y}. \quad [3.76]$$

Lastly, let us note, in the perspective of the preceding remark, that the molecular terms are most often neglected before the turbulent terms:

$$\nu \frac{\partial^2 \bar{u}}{\partial y^2} << \frac{\partial}{\partial y}\left(\overline{u'v'}\right) \quad [3.77]$$

$$\nu \frac{\partial^2 \bar{T}}{\partial y^2} << \frac{\partial}{\partial y}\left(\overline{T'v'}\right) \quad [3.78]$$

In the domain of inner boundary layers, the problem is most often solved empirically. Since the establishment regime is not predominant, Moody diagrams and a Darcy-type approach will be used to determine the friction (see **Chapter 5**).

3.3.2.2. *Empirical results*

Dynamic boundary layer

The values often used for the boundary layer thickness and friction on a flat plate are deduced from the value of the coefficient of friction:

$$\frac{\delta}{x} = \frac{0.382}{R_x^{0.2}} \quad [3.79]$$

$$C_f = \frac{0.0594}{R_x^{0.2}} \quad [3.80]$$

where $R_x = \dfrac{U_e\, x}{\nu}$ is the Reynolds number, based on the abscissa.

Thermal boundary layer

The thermal boundary layer thickness is often related to the dynamic boundary layer thickness, by a power of the Prandtl number:

$$\frac{\delta_{th}(x)}{\delta(x)} = \text{Pr}^{-1/3} \quad [3.81]$$

This result can be interpreted: the case of a very large number of Prandtl is, for example, that of liquid materials. This is how sodium, a metal that melts at a moderate temperature, is sometimes used for transfers within needles in nuclear technology (which is not without its own safety problems, due to the risk of fire)

In this case, the temperature gradient at the wall will be very high and the thickness of the thermal boundary layer will be very low.

The Nusselt and Stanton numbers are given by:

Values of the local transfer coefficient (Nusselt):

$$Nu_x = 0.0288 \quad R_x^{0.8} \quad P_r^{\frac{1}{3}} \quad [3.82]$$

Values of the average transfer coefficient (Nusselt) on L:

$$Nu_L = 0.036\, R_L^{0.8}\, P_r^{\frac{1}{3}} \quad [3.83]$$

We know that:

$$St = Nu_D\, R_D^{-1}\, \text{Pr} \quad [3.84]$$

We deduce the local Stanton number:

$$St_x = 0.0288 R_x^{-0.2} \quad P_r^{-2/3} \quad [3.85]$$

And the average Stanton number:

$$Nu_L = 0.036\, R_L^{-0.2}\, P_r^{-2/3} \qquad [3.86]$$

3.4. Problem of scale

The expressions here for the transfer coefficients are strictly valid when the wall temperature is constant. When the wall temperature is variable, the careless application of these formulas can lead to significant errors. For example, the application of the expression obtained in the laminar regime to a heating system with constant flow density can lead to 40%, or even 45%, of errors.

The literature thus contains many other studies, which we do not report here.

A particularly interesting case is the temperature step. This situation occurs when the heat transfer starts at a distance from the leading edge. Different expressions have been proposed to relate the local transfer, at an abscissa, to the convection coefficient (or Nusselt) obtained when the transfer at constant temperature starts at the leading edge.

The following expressions of the Stanton number are used for laminar and turbulent flows.

Eckert

This expression is valid in a laminar regime. It can be obtained by an integral method:

$$St_x = 0.0296 R_x^{0.8} \quad P_r^{1/3} \frac{1}{\left[1-\left(\frac{x}{x_0}\right)^{3/4}\right]^{1/3}} \qquad [3.87]$$

Rubesin

This expression is the result of a rather complex approach (the original report is about 50 pages long), based on an integral method. The coefficient may make some laugh today, but it results from this theory and we have retained it:

$$St_x = 0.0296 R_x^{0.8} \quad P_r^{1/3} \frac{1}{\left[1-\left(\frac{x}{x_0}\right)^{39/40}\right]^{7/39}} \qquad [3.88]$$

Seban

This expression has an experimental origin:

$$St_x = 0.0276 R_x^{0.8} \quad P_r^{1/9} \frac{1}{\left[1-\left(\frac{x}{x_0}\right)^{9/10}\right]^{1/9}} \qquad [3.89]$$

It is often admitted that Rubesin's formula is more adapted to gaseous flows (in terms of incompressible limitation), and Seban's formula to liquids.

Superposition calculation

The above expressions can be put in the following form:

$$St_x = g\left(\frac{x}{x_0}\right) St_{x,x_0=0} \qquad [3.90]$$

where $g\left(\frac{x}{x_0}\right)$ depends on the flow regime and the author proposing the formula.

x_0 is the abscissa of the beginning of the transfer and $St_{x,x_0=0}$ is the transfer without scale.

Lighthill proposed a clever method [LIG 50] to determine the heat transfer for a prescribed parietal temperature distribution $T_W(x)$:

$$St_x = \int_0^x g\left(\frac{x}{\xi}\right) St_{x,x_0=0} \, dT_W(\xi) \qquad [3.91]$$

In spirit, this method is equivalent to summing up the flows, at a point x, that result from a continuous sequence of small increments in temperature $dT_W(\xi)$, located at abscissas ξ distributed between 0 and x.

The calculation is thus numerical, but the transfer obtained will be theoretically accurate.

Note that from the mathematical point of view, $dT_W(\xi)$ is the differential of the curve $T_W(\xi)$ and that the resulting integral is a Stieljes integral.

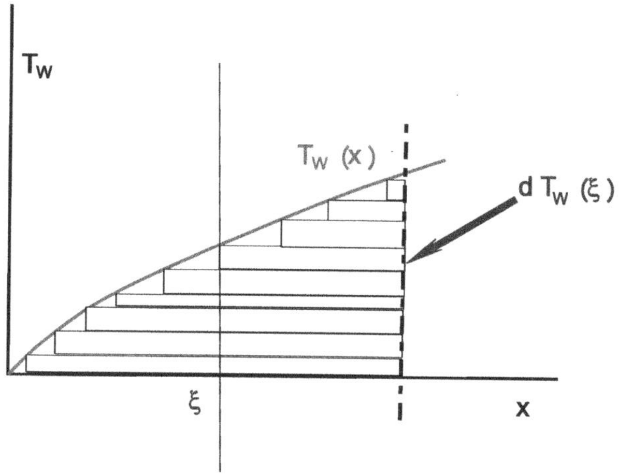

Figure 3.2. *Method: the parietal temperature at an abscissa x results from a series of small increments $dT_W(\xi)$ located at different abscissas ξ. For a color version of this figure, see www.iste.co.uk/ledoux/heat3.zip*

In the past, many semi-empirical methods have been tried for the treatment of turbulent flows. Today, the numerical approach is predominant, and for many industrial problems it is unavoidable.

3.5. Applications of the boundary layer theory

EXAMPLE 3.1.– Assessment of the boundary layer thickness at very low Prandtl numbers

In this case, it is known that the ratio of thermal and dynamic boundary layers $\frac{\delta_{th}}{\delta}$ is very strong. We generally give the following type of ratio: $\frac{\delta_{th}}{\delta} \approx \text{Pr}^{-0.5}$, where Pr is the Prandtl number.

Justify this expression with a simple model.

SOLUTION TO EXAMPLE 3.1.–

Although the thermal boundary layer is very large, almost all thermal phenomena occur in a uniform flow, identical to the external flow.

In this case:

$$u \approx U_e \qquad [3.92]$$

$$v \approx 0 \qquad [3.93]$$

The boundary layer equation becomes:

$$U_e \frac{\partial T}{\partial x} = \nu \frac{\partial^2 T}{\partial y^2} \qquad [3.94]$$

which is formally identical to the heat equation in a transient regime.

This is understandable, given that the conductive transfer occurs in a "block" of fluid moving at uniform speed. $\frac{U_e}{x}$ has the aspect of a time, over which the temperature evolves as a function of y.

The boundary conditions are:

$$y = 0 \quad ; \quad T = T_W \qquad [3.95]$$

$$y \to \infty \quad ; \quad T \to T_e$$

We can introduce the variable $\theta = \frac{T - T_W}{T_e - T_W}$; the system becomes:

$$U_e \frac{\partial \theta}{\partial x} = a \frac{\partial^2 \theta}{\partial y^2} \qquad [3.96]$$

$$y = 0 \quad ; \quad \theta = 0 \qquad [3.97]$$

$$y \to \infty \quad ; \quad \theta \to 1 \qquad [3.98]$$

The solution is based on the classical unsteady conduction equation:

Defining:

$$\eta = \sqrt{\frac{U_e y^2}{4ax}} \qquad [3.99]$$

$$\theta = erf\,\eta \qquad [3.100]$$

the thickness $y = \delta_{th}$ will be found for:

$$\theta = erf\,\eta \qquad [3.101]$$

$$\theta = 0.99 \qquad [3.102]$$

so: $\eta = 1.8$

Therefore:

$$1.9 = \sqrt{\frac{U_e \delta_{th}^2}{4ax}} \qquad [3.103]$$

From this, we deduce:

$$\delta_{th} = 3.8\sqrt{\frac{ax}{U_e}} \qquad [3.104]$$

and its relationship to the dynamic boundary layer:

$$\delta = 4.92\sqrt{\frac{\nu x}{U_e}} \qquad [3.105]$$

$$\frac{\delta_{th}}{\delta} = \frac{3.8}{4.92}\sqrt{\frac{a}{\nu}} = \frac{0.77}{\sqrt{\Pr}} \qquad [3.106]$$

This results in:

$$\frac{\delta_{th}}{\delta} \approx \Pr^{-0.5} \qquad [3.107]$$

EXAMPLE 3.2.– Laminar boundary layer with large Prandtl number

For large Prandtl numbers, it is known that the thermal boundary layer is much thinner than the dynamic boundary layer. The thermal phenomena therefore always “see” a virtually linear velocity profile.

We will rely on a Pohlhausen type of analysis. We assume that the results of the Blasius analysis are known, and particularly the functions f, f' and f''.

Let us recall that the velocity and temperature profiles are derived from two equations with boundary conditions.

We define a function $f(\eta)=\frac{\Psi}{\sqrt{\nu U_e x}}$ such that $f'(\eta)=\frac{u}{U_e}$.

It is shown that $f(\eta)$ answers to:

$$2f_{\eta\eta\eta}+ff''=0 \quad [3.108]$$

with:

$$f(0)=f'(0)=0 \quad [3.109]$$

$$f(\infty)\to 1 \quad [3.110]$$

and a reduced temperature function:

$$\theta(\eta)=\frac{T-T_W}{T_e-T_W} \quad [3.111]$$

This answers the so-called Pohlhausen equation:

$$2\theta''+\Pr f\theta'=0 \quad [3.112]$$

with:

$$\theta(0)=0 \quad [3.113]$$

$$\theta(\infty)\to 1 \quad [3.114]$$

1) For a linear profile (area very close to the wall, where thermal phenomena are confined), give the (thus approximate) expression of $\frac{u}{U_e}$ as a function of $f''(0)$.

2) Recalling the relationship of $\frac{u}{U_e}$ as a function of f', find the (approximate) expression of f.

3) Transferring this expression into the Pohlhausen equation, which has no reason to change form, find the expression of $\theta(\eta)$, by associating the boundary conditions.

4) Deduce the heat flow density, the expression of the Nusselt Nu_x, the Stanton St_x and their relation to the Reynolds R_x and the Prandtl Pr.

Any comments?

Appendix

Some of the following data may make the calculations easier.

We recall the definition of the "Gamma function":

$$\Gamma(x) = \int_0^{\infty} t^{x-1} e^{-t}\, dt \qquad [3.115]$$

We give:

$$\Gamma\left(\frac{1}{3}\right) = 2.679 \qquad [3.116]$$

$$\Gamma\left(\frac{1}{2}\right) = 1.772 \qquad [3.117]$$

$$\Gamma(1) = 1 \qquad [3.118]$$

We recall the recurrence relation:

$$\Gamma(x+1) = x\Gamma(x) \qquad [3.119]$$

Note that if x is an integer:

$$\Gamma(x) = (x-1)! \qquad [3.120]$$

The Gamma function is an extension to real numbers of the factorial function.

SOLUTION TO EXAMPLE 3.2.–

1) For a linear profile (area very close to the wall, where thermal phenomena are confined), give the expression (thus, approximate) of $\frac{u}{U_e}$ as a function of $f''(0)$.

1) If the velocity profile is linear, then:

$$\left.\frac{\partial u}{\partial y}\right|_y = \left.\frac{\partial u}{\partial y}\right|_{y=0} = \frac{\tau_W}{\mu} \qquad [3.121]$$

However:

$$f'(\eta) = \frac{u}{U_e} \qquad [3.122]$$

where:

$$\eta = y\sqrt{\frac{U_e}{\nu x}} \qquad [3.123]$$

Passing from (x, y) system to (ξ, η) system, we have:

$$\left.\frac{\partial u}{\partial y}\right|_y = \left.\frac{\partial U_e f'}{\partial y}\right|_\eta \qquad [3.124]$$

$$\left.\frac{\partial U_e f'}{\partial y}\right|_\eta = \sqrt{\frac{U_e}{\nu x}}\left.\frac{\partial U_e f'}{\partial \eta}\right|_\eta = \sqrt{\frac{U_e}{\nu x}}U_e f''(\eta) \qquad [3.125]$$

because: $\left.\dfrac{\partial U_e f'}{\partial \xi}\right|_{\eta} = 0$ [3.126]

In this case:

$$\left.\frac{\partial u}{\partial y}\right|_{y} = \left.\frac{\partial u}{\partial y}\right|_{y=0} = \sqrt{\frac{U_e}{\nu x}} U_e f''(0) \qquad [3.127]$$

$$u = \sqrt{\frac{\nu}{U_e x}} U_e f''(0) y = \eta f''(0) U_e \qquad [3.128a]$$

and $\dfrac{u}{U_e}$ is written as:

$$\frac{u}{U_e} = \eta f''(0) \qquad [3.128b]$$

2) Recalling the relationship of $\dfrac{u}{U_e}$ as a function of f', find the (approximate) expression of f.

Blasius' theory allowed us to write:

$$\frac{u}{U_e} = f'(\eta) \qquad [3.129]$$

Therefore, it can be deduced from a simple integration:

$$\frac{u}{U_e} = f' = \eta f''(0) \qquad [3.130]$$

$f''(0) =$ is a constant value, therefore:

$$f = \frac{\eta^2}{2} f''(0) + C \qquad [3.131]$$

$$f(0)=0 \quad ; \quad f''(0)=0.332\text{, so:}$$

$$f=0.166\eta^2 \qquad [3.132]$$

3) Transferring this expression into the Pohlhausen equation, which has no reason to change form, find the expression of $\theta(\eta)$, by associating the boundary conditions.

The Pohlhausen equation is usually written as:

$$2\theta''+\Pr f\theta'=0 \qquad [3.133]$$

In the present approximation, it becomes:

$$2\theta''+0.166\eta^2\Pr\theta'=0 \qquad [3.134]$$

which we can integrate to get an expression for θ' and θ'':

$$\frac{\theta''}{\theta'}=-\frac{0.166}{2}\eta^2\Pr \qquad [3.135]$$

$$Ln\,\theta'=-\frac{0.166}{6}\eta^3\Pr+Ln\,C_1=-2.77.10^{-2}\,\eta^3\Pr+Ln\,C_1 \qquad [3.136]$$

$$\theta'=C_1\,exp-2.77.10^{-2}\,\eta^3\Pr \qquad [3.137]$$

$$\theta=C_1\int_0^{\eta}exp\left(-2.77.10^{-2}\Pr\eta^3\right)d\bar{\eta}+C_2 \qquad [3.138]$$

Using the boundary conditions:

$$\theta(0)=0 \quad ; \quad C_2=0 \qquad [3.139]$$

$$\theta(\infty)\to 1 \quad ; \quad C_1=\frac{1}{\int_0^{\infty}exp\left(-2.77.10^{-2}\Pr\eta^3\right)d\bar{\eta}} \qquad [3.140]$$

So:

$$\theta(\eta) = \frac{\int_0^{\eta} exp\left(-2.77.10^{-2}\,\mathrm{Pr}\,\eta^3\right) d\bar{\eta}}{\int_0^{\infty} exp\left(-2.77.10^{-2}\,\mathrm{Pr}\,\eta^3\right) d\bar{\eta}} \quad [3.141]$$

We can evaluate the integral by laying down the following:

$$\xi = 2.77.10^{-2}\,\mathrm{Pr}\,\eta^3 \quad [3.142]$$

$$d\xi = 3 * 2.77.10^{-2}\,\mathrm{Pr}\,\eta^2\,d\eta \quad [3.143]$$

$$\eta = \left[\frac{\xi}{2.77.10^{-2}\,\mathrm{Pr}}\right]^{\frac{1}{3}} \quad [3.144]$$

$$d\xi = 3 * 2.77.10^{-2}\,\mathrm{Pr}\,\eta^2\,d\eta \quad [3.145]$$

$$d\eta = \frac{d\xi}{3 * 2.77.10^{-2}\,\mathrm{Pr}\,\eta^2} = \frac{d\xi}{3 * 2.77.10^{-2}\,\mathrm{Pr}\left[\frac{\xi}{2.77.10^{-2}\,\mathrm{Pr}}\right]^{\frac{2}{3}}} = \frac{\xi^{-\frac{2}{3}}\,d\xi}{3 * \left(2.77.10^{-2}\,\mathrm{Pr}\right)^{\frac{1}{3}}}$$

$$\int_0^{\infty} exp\left(-2.77.10^{-2}\,\mathrm{Pr}\,\eta^3\right) d\bar{\eta} = \int_0^{\infty} \frac{\xi^{-\frac{2}{3}}}{3 * \left(2.77.10^{-2}\,\mathrm{Pr}\right)^{\frac{1}{3}}}\,exp(-\xi)\,d\xi$$

$$\int_0^{\infty} exp\left(-2.77.10^{-2}\,\mathrm{Pr}\,\eta^3\right) d\bar{\eta} = 1.101 \int_0^{\infty} \xi^{\frac{1}{3}-1}\,exp(-\xi)\,d\xi = 1.101\,\mathrm{Pr}^{-\frac{1}{3}}\,\Gamma\left(\frac{1}{3}\right)$$

$$\Gamma\left(\frac{1}{3}\right) = 2.679$$

$$\frac{1}{\int_0^{\infty} exp\left(-2.77.10^{-2}\,\mathrm{Pr}\,\eta^3\right) d\bar{\eta}} = 0.339.\mathrm{Pr}^{\frac{1}{3}}$$

$$\theta'(\eta) = \frac{exp\left(-2.77.10^{-2}\,\mathrm{Pr}\,\eta^3\right)}{\int_0^{\infty} exp\left(-2.77.10^{-2}\,\mathrm{Pr}\,\eta^3\right) d\bar{\eta}} \quad [3.146]$$

$$\theta'(0) = \frac{1}{\int_0^{\infty} exp\left(-2.77.10^{-2}\,\mathrm{Pr}\,\eta^3\right) d\bar{\eta}} = 0.339\,\mathrm{Pr}^{\frac{1}{3}} \quad [3.147]$$

$$\varphi_W = \lambda \left.\frac{\partial T}{\partial y}\right|_{y=0} = \sqrt{\frac{U_e}{\nu x}}\,\lambda \left.\frac{\partial T}{\partial \eta}\right|_{\eta=0} = \sqrt{\frac{U_e}{\nu x}}\,\lambda\left(T_e - T_W\right) \left.\frac{\partial \theta}{\partial \eta}\right|_{\eta=0} \quad [3.148]$$

$$\varphi_W = \sqrt{\frac{U_e}{\nu x}}\,\lambda\left(T_e - T_W\right)\theta'(0)\,\Big| = 0.339\,\mathrm{Pr}^{\frac{1}{3}} \sqrt{\frac{U_e}{\nu x}}\,\lambda\left(T_e - T_W\right)\Big| \quad [3.149]$$

4) From the heat flow density φ_W, we can directly deduce the expression of the Nusselt Nu_x, the Stanton St_x and their relationship to the Reynolds R_x and the Prandtl Pr.

$$Nu_x = \frac{\varphi_W\, x}{\lambda\left(T_e - T_W\right)} = 0.339\,\mathrm{Pr}^{\frac{1}{3}} \sqrt{\frac{U_e}{\nu x}}\, x\Big| \quad [3.150]$$

$$Nu_x = 0.339\,\mathrm{Pr}^{\frac{1}{3}}\, R_x^{\frac{1}{2}} \quad [3.151]$$

This result is a very good approximation of the already-known formula:

$$Nu_x = 0.332\,\mathrm{Pr}^{\frac{1}{3}}\, R_x^{\frac{1}{2}} \quad [3.152]$$

Similarly, the present model gives the Stanton number:

$$St_x = \frac{Nu_x}{\mathrm{Pr}\, R_x} = 0.339\,\mathrm{Pr}^{-\frac{1}{3}}\, R_x^{-\frac{1}{2}} \quad [3.153]$$

EXAMPLE 3.3.– Plate in the path of the wind. Friction and heat transfer with parietal fluid extraction.

We intend to study the boundary layer produced by a uniform flow with the distanced velocity U_e and a porous flat plate placed in the path of the wind, parallel to the current lines of the potential flow. The fluid is incompressible, of density ρ, dynamic viscosity μ. All these physical properties are constant.

Fluid is extracted along the entire length of the plate with a constant velocity of modulus W.

The flow, being a plane flow, will be marked by a system of axes Oxy (following the plate). The origin O is chosen at the leading edge. The velocity will have two components u and v respectively along Ox and Oy.

We will assume that from a certain abscissa x_0, the flow is self-similar, that is, the longitudinal component of the local velocity only depends on the distance to the wall, y: $u = u(y)$.

The area near the leading edge ($x<x_0$) will be called the non-similar area.

We will solve the problem using an analytical method. We will propose an approach using an integral method in section 3.2.

1) From the continuity writing, deduce the profile of the lateral component v of the velocity in the boundary layer.

2) Deduce the simple form that the momentum equation then takes.

3) Determination of velocity and friction profiles.

a) Find the velocity profile $u(y)$.

b) Deduce the dynamic boundary layer thickness from this.

c) Give the expressions for the tangential parietal stress τ_W and the coefficient of friction C_f.

4) Determination of reduced temperature profiles and parietal heat transfer.

a) We introduce the reduced thermal variable:

$$\theta(y) = \frac{T - T_W}{T_e - T_W} \quad [3.154]$$

It is assumed that in the area of similarity, $\theta = \frac{T - T_W}{T_e - T_W}$ only varies with y: $\theta = \theta(y)$.

Find the shape of the profile $\theta(y)$.

Express it as a function of the Prandtl number of the fluid P_r and δ.

b) Deduce the parietal heat flow $\varphi_W(x)$ and the Stanton number $S_t(x)$.

c) Could we not have found $\varphi_W(x)$ and $S_t(x)$ directly without all this calculation?

SOLUTION TO EXAMPLE 3.3.–

1) Profile of $v(y)$

For a two-dimensional flow, which is plane and incompressible, the continuity equation is written as:

$$\frac{\partial u}{\partial x} + \frac{\partial v}{\partial y} = 0 \quad [3.155]$$

It is reduced here to:

$$u = u(y) \; ; \; \frac{\partial u}{\partial x} = 0 \quad [3.156]$$

$$\frac{\partial v}{\partial y} = 0 \quad [3.157]$$

v is constant in the boundary layer, and will therefore have the value set at the wall:

$$v = -W \quad [3.158]$$

2) The momentum equation of the two-dimensional boundary layer, which is plane and incompressible, then simplifies to:

$$u\frac{\partial u}{\partial x}+v\frac{\partial u}{\partial y}=-\frac{1}{\rho}\frac{\partial p}{\partial x}+\nu\frac{\partial^2 u}{\partial y^2} \qquad [3.159]$$

In the absence of a longitudinal gradient of external velocity:

$$-\frac{1}{\rho}\frac{\partial p}{\partial x}=U_e\frac{dU_e}{dx}=0 \qquad [3.160]$$

The final shape emerges:

$$0-W\frac{\partial u}{\partial y}=\nu\frac{\partial^2 u}{\partial y^2} \qquad [3.161]$$

$$y=0\ ;\ u=0 \qquad [3.162]$$

a) Velocity profile

The equation in $u(y)$ is easily integrated:

$$0-W\frac{\partial u}{\partial y}=\nu\frac{\partial^2 u}{\partial y^2} \qquad [3.163]$$

$$\frac{\dfrac{\partial^2 u}{\partial y^2}}{\dfrac{\partial u}{\partial y}}=-\frac{W}{\nu} \qquad [3.164]$$

$$Ln\left(\frac{\partial u}{\partial y}\right)=-\frac{W}{\nu}y+Ln\,C_1 \qquad [3.165]$$

Taking the exponential of the two terms:

$$\frac{\partial u}{\partial y}=C_1\exp-\frac{W}{\nu}y \qquad [3.166]$$

$$u = \int_0^y C_1 \exp - \frac{W}{\nu} dy + C_2 \qquad [3.167]$$

$$u = -\frac{\nu}{W} C_1 \exp - \frac{W}{\nu} y \Big|_0^y + C_2 = -\frac{\nu}{W} C_1 \left(\exp - \frac{W}{\nu} y \right) - 1 + C_2 \qquad [3.168]$$

Note that we could also have written $u = \int_{C_2}^{y} C_1 \exp - \frac{W}{\nu} dy$, which obviously leads to the same final result.

By applying the boundary conditions:

$$y = 0 \; ; \; u = 0 \; ; \; C_2 = 0 \qquad [3.169]$$

$$y \to \infty \; ; \; u \to U_e ; \quad U_e = -\frac{\nu}{W} C_1 \; (0 - 1) + C_2 = \frac{\nu}{W} C_1 ; \quad C_1 = \frac{W U_e}{\nu} \qquad [3.170]$$

Finally:

$$u = U_e \left(1 - \exp - \frac{W}{\nu} y \right) \qquad [3.171]$$

Note that $\frac{W}{\nu} y$ is a Reynolds number.

b) Boundary layer thickness

By definition of the boundary layer border:

$$y = \delta \; ; \; \frac{u}{U_e} = 0.99 \qquad [3.172]$$

$$1 - \exp - \frac{W}{\nu} \delta = 0.99 \qquad [3.173]$$

$$\delta = 4.605 \frac{\nu}{W} \qquad [3.174]$$

So, by expressing the result in dimensionless form through a Reynolds:

$$\frac{W \delta}{\nu} = 4.605 \qquad [3.175]$$

We note that in the self-similar flow region, the boundary layer thickness is constant, as well as the Reynolds number $\frac{W\,\delta}{\nu}$.

c) Friction

The tangential parietal stress can be calculated directly from the velocity profile:

$$\tau = \mu \frac{du}{dy} = \mu \frac{d}{dy} U_e \left(1 - \exp - \frac{W}{\nu} y\right)_{y=0} = \rho W U_e \tag{3.176}$$

$$C_f = \frac{2\tau_W}{\rho U_e^2} = 2\frac{W}{U_e} \tag{3.177}$$

4) The heat transfer will be calculated by a similar calculation to the previous one.

a) Using the reasoning in 1.1 and 2.1, the energy equation is written as:

$$0 - W \frac{\partial T}{\partial y} = a \frac{\partial^2 T}{\partial y^2} \tag{3.178}$$

With boundary conditions:

$$y = 0 \;;\; T = T_W \tag{3.179}$$

$$y \to \infty \;;\; T \to T_e \tag{3.180}$$

which can be rewritten as:

$$0 - W \frac{\partial \theta}{\partial y} = a \frac{\partial^2 \theta}{\partial y^2} \tag{3.181}$$

With boundary conditions:

$$y = 0 \;;\; \theta = 0 \tag{3.182}$$

$$y \to \infty \;;\; \theta \to 1 \tag{3.183}$$

The equation in $\theta(y)$ is easily integrated:

$$0 - W\frac{\partial \theta}{\partial y} = a\frac{\partial^2 \theta}{\partial y^2} \quad [3.184]$$

$$\frac{\dfrac{\partial^2 \theta}{\partial y^2}}{\dfrac{\partial \theta}{\partial y}} = -\frac{W}{a} \quad [3.185]$$

$$Ln\left(\frac{\partial \theta}{\partial y}\right) = -\frac{W}{a}y + Ln\,C_1 \quad [3.186]$$

Taking the exponential of the two terms:

$$\frac{\partial \theta}{\partial y} = C_1 \exp -\frac{W}{a}y \quad [3.187]$$

$$\theta = \int_0^y C_1 \exp -\frac{W}{a}dy + C_2 \quad [3.188]$$

$$\theta = -\frac{a}{W}C_1 \exp -\frac{W}{a}y\Big|_0^y = -\frac{a}{W}C_1\left(\exp -\frac{W}{a}y\right) - 1 + C_2 \quad [3.189]$$

Note that we could also have written $\theta = \int_{C_2}^{y} C_1 \exp -\frac{W}{a}dy$, which obviously leads to the same final result.

By applying the boundary conditions:

$$y = 0 \;;\; \theta = 0 \quad ; \quad C_2 = 0 \quad [3.190]$$

$$y \to \infty \;;\; \theta \to 1 \quad ; \quad 1 = -\frac{a}{W}C_1\left(0 - 1|\right) + C_2 = \frac{a}{W}C_1 \quad ; \quad C_1 = \frac{W}{a} \quad [3.191]$$

Finally:

$$\theta = 1 - \exp -\frac{W}{a}y \quad [3.192]$$

b) The flow will be deduced from this:

$$\varphi = \lambda \frac{\partial T}{\partial y}\bigg|_{y=0} = \lambda(T_W - T_e)\frac{\partial \theta}{\partial y}\bigg|_{y=0} = \lambda(T_W - T_e)\frac{\partial}{\partial y}\left(1 - \exp - \frac{W}{a} y\right)\bigg|_{y=0} \quad [3.193]$$

$$\varphi = \lambda(T_W - T_e)\left(\frac{W}{a}\exp - \frac{W}{a} y\right)\bigg|_{y=0} = \lambda(T_W - T_e)\frac{W}{a} \quad [3.194]$$

and the Nusselt and Stanton numbers are easily deduced:

$$Nu(x) = \frac{\varphi x}{\lambda(T_W - T_e)} = \frac{W x}{a} \quad [3.195]$$

$$St(x) = \frac{Nu(x)}{R_x \Pr} = \frac{W x}{a \dfrac{U_e x}{\nu} \dfrac{\nu}{a}} \quad [3.196]$$

$$St(x) = \frac{W}{U_e} \quad [3.197]$$

The thermal boundary layer thickness δ_{th} is:

$$\theta = 1 - \exp - \frac{W}{a}\delta_{th} = 0.99 \quad [3.198]$$

$$\delta_{th} = 4.605 \frac{a}{W} \quad [3.199]$$

Remembering that: $\delta = 4.605 \dfrac{\nu}{W}$

We see that: $\dfrac{\delta_{th}}{\delta} = \dfrac{a}{\nu} = \Pr^{-1}$ [3.200]

c) We could have just written that the heat exchanged per unit of surface area between the plate and the gas was given by the enthalpy flow, allowing the velocity gas to go from T_e to T_w:

$$\varphi = \rho c_P \left(T_W - T_e\right) W = \frac{\lambda \rho c_P}{\lambda} \left(T_W - T_e\right) W = \lambda \left(T_W - T_e\right) \frac{W}{a} \qquad [3.201]$$

EXAMPLE 3.4.– Thermal transfer in a falling film

Although the objective of this exercise is to calculate a heat flow to a vertical film, we present the following dynamic problem, for educational reasons, from the perspective of an inclined film.

We consider the flow in an inclined channel in relation to the horizontal by an angle. μ=1.8.10^{-3} *Pl*. In this channel, we assume that the flow is strictly parallel and that its depth h is constant. Let $\mu = 1.8.10^{-3}\,Pl$ be the viscosity of water and $\rho = 1000\ kg.m^{-3}$ its density. Air, on the other hand, will be considered as a perfect fluid.

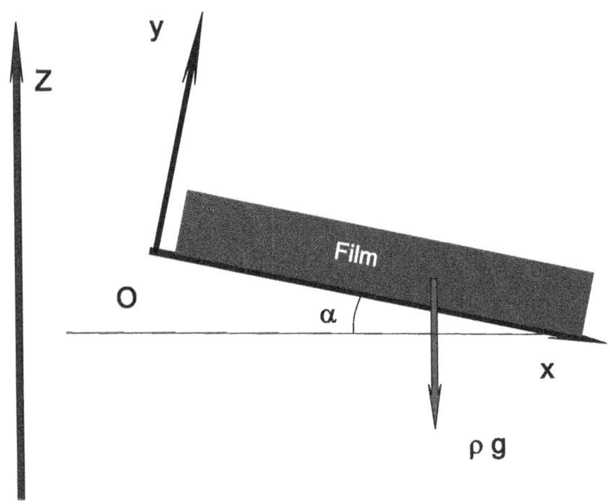

Figure 3.3. *Inclined film: the coordinate system. For a color version of this figure, see www.iste.co.uk/ledoux/heat3.zip*

1) The system is reduced to a reference frame *Ox*, *Oy*, where *Ox* is located on the bottom of the channel.

1.1) Show that the velocity $\vec{V}(u, v, w)$ is reduced to a single component *u*, that is only variable with *y*.

1.2) Give the form of the variation of the generating pressure $p_G(x)$.

1.3) Write the differential equation which obeys $u(y)$.

1.4) Give the expression of $u(y)$.

1.5) Give the expression of the volume flow q_v in the channel, in literal form.

2) We now consider a vertical film of water, of constant thickness δ and width.

What happens to the results of the previous question?

3) The vertical wall is heated to a temperature T_W. The air outside the film is at temperature T_e.

It is assumed that a thermal zone is "established" fairly quickly, where the temperature no longer depends on the altitude.

In this established area, determine the main elements of heat transfer:

– temperature profile;

– parietal flux density φ_W;

– Nusselt number Nu_δ related to the film thickness;

– Stanton number.

It is useful to define the dimensionless variable θ:

$$\theta = \frac{T - T_W}{T_e - T_W} \quad [3.202]$$

SOLUTION TO EXAMPLE 3.4.–

1) The flow satisfies the continuity equation.

1.1) It must be written in two dimensions, for an incompressible fluid:

$$div\vec{V} = 0 \quad ; \quad \frac{\partial u}{\partial x} + \frac{\partial v}{\partial y} = 0$$

We have assumed a parallel film flow. Therefore, the velocity reduces to its component that is parallel to the wall, u, which implies $v=0$. As a result:

$$\frac{\partial u}{\partial x}+0=0 \qquad [3.203]$$

and $u=u(y)$ is now only a function of y.

1.2) Let us define a vertical axis OZ. This axis must be distinguished from a OZ axis, which is not shown here. This would end the orthonormal (Ox, Oy, Oz) reference frame.

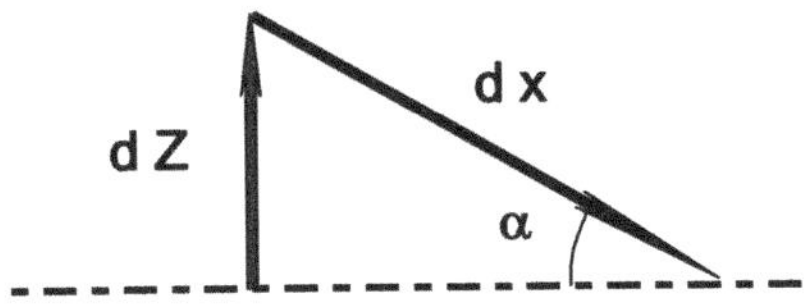

Figure 3.4. *Relation of dz to dx. For a color version of this figure, see www.iste.co.uk/ledoux/heat3.zip*

OZ and Oy obviously make an angle. α is the angle that the channel makes with respect to the horizontal plane. It can be easily calculated, knowing that the channel descends 12.5 m over 50 km:

$$\sin\alpha=\frac{12.5}{50000}=2.5.10^{-4} \quad ; \quad \alpha=1.43.10^{-2\circ} \qquad [3.204]$$

Therefore: $dZ=-dx\sin\alpha$

The x derivative of the generating pressure will then be written as:

$$\frac{d\,p_G}{d\,x}=\frac{d(p+\rho g Z)}{d\,x}=\frac{d\,p}{d\,x}+\rho g\frac{d\,Z}{d\,x}=\frac{d\,p}{d\,x}-\rho g\sin\alpha \qquad [3.205]$$

1.3) The equation of dynamics, or the momentum equation, for a two-dimensional flow of an incompressible Newtonian fluid is written as:

$$\nu\left[\frac{\partial^2 u}{\partial y^2}\right]+\rho g\sin\alpha=0 \qquad [3.206]$$

u only depends on y, the partial derivatives can be transformed into "right" derivatives:

$$\nu\left[\frac{d^2u}{d\,y^2}\right]+\rho g\sin\alpha=0 \qquad [3.207]$$

The forces of volume reduced to gravity, brought back in a reference frame of axis $0x$ making an angle α with the horizontal plane, will be reduced to:

$$F_{Vx}=\rho g\sin\alpha \qquad [3.208]$$

Given that $\dfrac{\partial u}{\partial x}$ and v are both zero, the equation of dynamics is finally reduced to two equivalent entries:

$$-\frac{1}{\rho}\frac{\partial p}{\partial x}+\nu\left[\frac{d^2u}{d\,y^2}\right]+\rho g\sin\alpha=0 \qquad [3.209]$$

or:

$$-\frac{1}{\rho}\frac{\partial p_G}{\partial x}+\nu\left[\frac{d^2u}{d\,y^2}\right]=0 \qquad [3.210]$$

In any plane that is perpendicular to the current lines of the film, the generating pressure p_G is constant. Thus, for any abscissa x, it is equal to its value on the free surface. At this point, the static pressure is equal to the atmospheric pressure p_a. We therefore have:

$$\frac{d\,p_G}{d\,x}=\frac{d\,p}{d\,x}-\rho g\sin\alpha=\frac{d\,p_a}{d\,x}-\rho g\sin\alpha=0-\rho g\sin\alpha \qquad [3.211]$$

The momentum equation is therefore written as:

$$\mu\left[\frac{d^2u}{d\,y^2}\right]+\rho g\sin\alpha=0 \qquad [3.212]$$

The boundary conditions will simply be written as:

a) No-slip condition: at the bottom of the channel, the speed is zero:

$$y=0 \ ; \ u=0. \qquad [3.213]$$

b) The air is a perfect fluid: the tangential stress it applies on the surface of the flow.

Due to the law of action and reaction, the tangential stress in the water on the free surface is zero:

$$y=h \ ; \ \mu\frac{du}{dy}=0 \qquad [3.214]$$

1.4) The solution for the second-degree linear equation is simple. It proceeds through a double integration, in which y will give two constants, C_1 and C_2, that we will determine by the boundary conditions.

$$\mu\left[\frac{d^2u}{dy^2}\right]+\rho g\sin\alpha=0 \qquad [3.215]$$

$$\frac{d^2u}{dy^2} = -\frac{g}{\nu}\sin\alpha=0 \qquad [3.216]$$

Let us integrate a first time:

$$\frac{du}{dy} = -\left(\frac{g}{\nu}\sin\alpha\right)y+C_1 \qquad [3.217]$$

Let us integrate a second time:

$$u = -\left(\frac{g}{\nu}\sin\alpha\right)\frac{y^2}{2}+C_1y+C_2 \qquad [3.218]$$

Let us now apply the boundary conditions:

$$y=0 \ ; \ u=0= -\left(\frac{g}{\nu}\sin\alpha\right)*\frac{0^2}{2}+C_1*0+C_2 \quad ; \quad C_2=0 \qquad [3.219]$$

$$y = h \;;\; \mu \frac{du}{dy} = 0 \;;\; \frac{du}{dy} = 0 = -\left(\frac{g}{\nu}\sin\alpha\right)h + C_1 \quad ; \quad C_1 = \left(\frac{g}{\nu}\sin\alpha\right)h$$

The expression of u:

$$u(y) = -\left(\frac{g}{\nu}\sin\alpha\right)\frac{y^2}{2} + \left(\frac{g\,h}{\nu}\sin\alpha\right)y \qquad [3.220]$$

This expression can be reformulated in a dimensionless form, showing a reduced velocity $\tilde{u} = \dfrac{u}{U_0}$ and a reduced ordinate $\xi = \dfrac{y}{h}$:

$$u(y) = -\left(\frac{g}{\nu}\sin\alpha\right)\frac{y^2}{2}\frac{h^2}{h^2} + \left(\frac{g\,h}{\nu}\sin\alpha\right)\; y\frac{h}{h} \qquad [3.221]$$

$$u(y) = U_0\left(-\frac{y^2}{2h^2} + \frac{y}{h}\right) = U_0\left(\xi - \frac{\xi^2}{2}\right) \qquad [3.222]$$

$$U_0 = \frac{g\,h^2}{\nu}\sin\alpha \qquad [3.223]$$

$$\tilde{u}(\xi) = \xi - \frac{\xi^2}{2} \qquad [3.224]$$

1.5) To calculate the volume flow q_v in the channel, we will divide a section into small horizontal strips of width l of the channel and height dy, therefore area $dS = l\,dy$, taking into account the variation of u with y. It can therefore be written in a classical way:

$$q_v = \int_0^h u(y)\,d\,S = \int_0^h u(y)\,l\,d\,y \qquad [3.225]$$

After a change of variable:

$$\xi = \frac{y}{h} \;;\; d\xi = \frac{d\,y}{h} \quad q_v = \int_0^h u(y)\,d\,S = \int_0^1 u(y)\,l\,h\,d\,\xi \qquad [3.226]$$

$$q_v = \int_0^1 u(y) l\, h\, d\xi = \int_0^1 U_0 \left(\xi - \frac{\xi^2}{2} \right) l\, h\, d\xi \quad [3.227]$$

$$U_0 = \frac{g\, h^2}{\nu} \sin \alpha \quad [3.228]$$

$$q_v = l\, h U_0 \int_0^1 \left(\xi - \frac{\xi^2}{2} \right) d\xi = l\, h U_0 \left. \frac{\xi^2}{2} - \frac{\xi 3}{6} \right|_0^1 \quad [3.239]$$

The final expression of the flow rate is:

$$q_v = \frac{l\, h U_0}{3} = \frac{g\, l\, h^3}{3\nu} \sin \alpha \quad [3.230]$$

The previous results remain valid for a vertical film. We will consider a width l equal to the unit.

It is enough to put $\alpha = \frac{\pi}{2}$, or $\sin \alpha = 1$; h is now equal to δ.

The expression of the velocity profile will be:

$$u(y) = \frac{g\, \delta}{\nu} y - \frac{g}{\nu} \frac{y^2}{2} \quad [3.231]$$

which we can write as:

$$u(y) = U_0 \frac{y}{\delta} - \frac{1}{2} \frac{y^2}{\delta^2} \quad [3.232]$$

with

$$U_0 = \frac{g\, \delta^2}{\nu} \quad [3.233]$$

The flow rate will be:

$$q_v = \frac{l\, \delta U_0}{3} = \frac{g\, l\, \delta^3}{3\nu} \quad [3.234]$$

3) Heat transfer in an established zone

In this parallel flow, the energy equation takes a particularly simple form:

$$u\frac{\partial T}{\partial x} = a\frac{\partial^2 T}{\partial y^2} \quad [3.235]$$

Let us use the dimensionless variable θ:

$$\theta = \frac{T - T_W}{T_e - T_W} \quad [3.236]$$

The general equation becomes:

$$u\frac{\partial \theta}{\partial x} = a\frac{\partial^2 \theta}{\partial y^2} \quad [3.237]$$

with boundary conditions:

$$y = 0 \quad ; \quad \theta = 0 \quad [3.238]$$

$$y = \delta \quad ; \quad \theta = 1 \quad [3.239]$$

In the established zone, we put: $\frac{\partial T}{\partial x} = 0$

The equation to solve becomes particularly simple:

$$a\frac{\partial^2 \theta}{\partial y^2} = 0 \quad [3.240]$$

with

$$y = 0 \quad ; \quad \theta = 0 \quad [3.241]$$

$$y = \delta \quad ; \quad \theta = 1 \quad [3.242]$$

The solution is obvious; it is a linear function. Using the boundary conditions, we immediately find:

$$\theta = \frac{y}{\delta} \quad [3.243]$$

Let us now note that:

$$\left.\frac{\partial \theta}{\partial y}\right|_{y=0} = \frac{\partial \theta}{\partial y} = \frac{1}{\delta} \qquad [3.244]$$

The heat flow density is deduced:

$$\varphi_W = \lambda \left.\frac{\partial T}{\partial y}\right|_{y=0} = \lambda\left(T_W - T_e\right)\left.\frac{\partial \theta}{\partial y}\right|_{y=0} = \frac{\lambda\left(T_W - T_e\right)}{\delta} \qquad [3.245]$$

The Nusselt will be:

$$Nu_\delta = \frac{h\delta}{\lambda} = \frac{\delta}{\lambda}\frac{\varphi_W}{T_W - T_e} = 1 \qquad [3.246]$$

and the Stanton will be:

$$St_\delta = \frac{\varphi_W}{\rho c_P U_0\left(T_W - T_e\right)} = \frac{\lambda\left(T_W - T_e\right)}{\rho c_P U_0\left(T_W - T_e\right)\delta} = \frac{a}{U_0\,\delta} \qquad [3.247]$$

We have chosen the maximum velocity U_0 as a reference. The Stanton appears here as the opposite of a Péclet number.

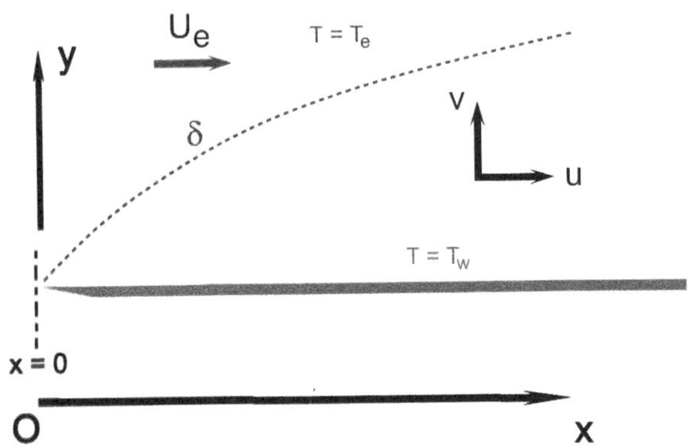

Figure 3.5. *Coordinates and notations. For a color version of this figure, see www.iste.co.uk/ledoux/heat3.zip*

EXAMPLE 3.5.– Kinetic heating

A uniform gas flow, at very high velocity U_e of temperature T_e, touches a perfectly adiabatic flat plate (no transfer is possible with the fluid), located parallel to its flow lines. This plate has a temperature T_w, which results from the studied phenomenon.

It will be assumed that the resulting boundary layer is laminar.

The theory developed here will be made under the assumption that, although the physical properties of the fluid are variable in the boundary layer, all the transfer takes place as if the properties were constant and calculated at a so-called reference temperature T_m:

$$T_m = \frac{T_e + T_w}{2} \quad [3.248]$$

1) Even in the absence of heat transfer to the wall, for sufficiently high U_e velocities, T_w is significantly greater than T_e. Explain the phenomenon qualitatively.

2) The recovery factor r is defined by the expression:

$$r = \frac{T_w - T_e}{U_e^2} 2c_p \quad [3.249]$$

where c_p is the mass heat capacity of the gas.

What is the physical meaning of this coefficient?

3) Putting it into equation and solving

3.1) Write the system of three equations in the usual Cartesian system (see figure), describing the phenomenon, as well as the boundary conditions.

3.2) Apply two transformations to this system of equations:

A transformation of coordinates from a (x, y) system to a (x, η) system:

(x, η):

$$\xi \equiv x \quad [3.250]$$

$$\eta = \frac{y}{x}\sqrt{R_x} \quad [3.251]$$

where R_x is the Reynolds number calculated with physical properties, determined at temperature T_m : $R_x = \frac{U_e x}{\nu}$.

A transformation of variables, through the definition of:

$$f(\eta) = \frac{\psi(x,y)}{\sqrt{\nu U_e x}} \quad [3.252]$$

$$\Xi(\eta) = \frac{T - T_w}{U_e^2} 2c_p \quad [3.253]$$

where $\psi\,(x, y)$ isthe current function of the flow.

We will assume that the affinity hypothesis is verified:

$$\frac{\partial f}{\partial x} = \frac{\partial \xi}{\partial x} = 0 \quad [3.254]$$

It should be noted that:

$$f' = \frac{\partial f}{\partial \eta} \quad [3.255]$$

$$f'' = \frac{\partial^2 f}{\partial \eta^2} \quad [3.256]$$

$$\xi' = \frac{\partial \xi}{\partial \eta} \quad [3.257]$$

Also write the expression in this system of boundary conditions. We will use Pr for the Prandtl number of the air at T_m.

3.3) What is the significance of $\Xi(0)$ when there may be a parietal heat flow density?

4) Solution to the heat equation.

4.1) Show that the solution for $\Xi(\eta)$ has the form:

$$\Xi(\eta) = 2\,\mathrm{Pr}\int_{\eta}^{\infty} f''(\bar{\eta})^{P_r} \left[\int_0^{\bar{\eta}} f''(\bar{\bar{\eta}})^{2-P_r}\, d\bar{\bar{\eta}}\right] d\bar{\eta} \qquad [3.258]$$

where $\bar{\eta}$ and $\bar{\bar{\eta}}$ are integration variables.

4.2) Give the expression for the parietal heat factor r as a function of $f(\eta)$ and the Prandtl number.

4.3) It has been shown in the literature that a good approximation of this factor is given by:

$$r \approx \sqrt{\mathrm{Pr}} \qquad [3.259]$$

Check it here.

SOLUTION TO EXAMPLE 3.5.–

1) Summary analysis of the phenomenon

When the velocity at infinity increases, the thickness $\delta(x)$ of the boundary layer decreases, the lateral velocity gradients increase, as does the friction. The dissipation of kinetic energy into thermal energy increases. At the measured velocity values, this heating remains insignificant. When the boundary layer becomes small, the thermalization by friction becomes notable and must be taken into account in the balances of boundary layer energy. The wall temperature will then be increased compared to the external flow temperature.

In Blasius and Pohlhausen's analysis, a term $\mu\left(\frac{\partial u}{\partial y}\right)^2$ is neglected. We will see that this term becomes important here.

2) The recovery factor r is defined by the expression:

$$r = \frac{T_w - T_e}{U_e^2} 2c_p \qquad [3.260]$$

where c_p is the mass heat capacity of the gas.

We can rewrite this expression as:

$$2c_p\left(T_w - T_e\right) = r\frac{U_e^2}{2} \qquad [3.261]$$

Note that r can be interpreted as the fraction of mass kinetic energy in the free flow required for parietal heating of the mass unit.

This coefficient r **thus makes it possible to calculate the heating of the wall (relative to the external flow)** according to the velocity of the external flow and the thermal properties of the fluid.

It is also worth noting that $\Xi(\eta) = \frac{T - T_w}{U_e^2} 2c_p$ can be related by r to $\theta(\eta) = \frac{T - T_w}{T_e - T_w}$, the function used in the traditional Pohlhausen theory:

$$\Xi(\eta) = \frac{T - T_w}{T_e - T_w} 2c_p \frac{T_e - T_w}{U_e^2} = r\theta \qquad [3.262]$$

3) Putting it into equation and solving

3.1) The system of three equations in the usual Cartesian system will show a kinetic term in the energy equation $\mu\left(\frac{\partial u}{\partial y}\right)^2$:

$$\frac{\partial u}{\partial x} + \frac{\partial v}{\partial y} = 0 \qquad [3.263]$$

$$u\frac{\partial u}{\partial x} + v\frac{\partial v}{\partial y} = -\frac{1}{\rho}\frac{d p_e}{d x} + \nu\frac{\partial^2 u}{\partial y^2} \qquad [3.264]$$

$$u\frac{\partial T}{\partial x}+v\frac{\partial T}{\partial y} = a\frac{\partial^2 T}{\partial y^2}+\mu\left(\frac{\partial u}{\partial y}\right)^2 \quad [3.265]$$

3.2) Let us apply the two pre-required transformations to this system of equations.

a) Regarding the momentum equation, the calculation established by Blasius remains unchanged. We will find:

The Blasius equation:

$$2f_{\eta\eta\eta}+ff''=0 \quad [3.266]$$

Assorted boundary conditions:

$$f(0)=0 \;;\; f'(0)=0 \;;\; f'(\infty)\to 1 \quad [3.267]$$

b) For the energy equation, the Pohlhausen equation will no longer be valid, because of the additional term:

$\dfrac{\mu}{\rho c_p}\left(\dfrac{\partial u}{\partial y}\right)^2$, which is the dissipation function after applying the boundary layer simplifications;

$$u\frac{\partial T}{\partial x}+v\frac{\partial T}{\partial y} = a\frac{\partial^2 T}{\partial y^2}+\frac{\mu}{\rho c_p}\left(\frac{\partial u}{\partial y}\right)^2 \quad [3.268]$$

Introducing the function $\Xi(\eta)=\dfrac{T-T_w}{U_e^2}2c_p$, the energy equation becomes:

$$u\frac{\partial \Xi}{\partial x}+v\frac{\partial \Xi}{\partial y} = a\frac{\partial^2 \Xi}{\partial y^2}+\frac{2\mu c_p}{\rho c_p U_e^2}\left(\frac{\partial u}{\partial y}\right)^2 \quad [3.269]$$

So: $$u\frac{\partial \Xi}{\partial x}+v\frac{\partial \Xi}{\partial y} = a\frac{\partial^2 \Xi}{\partial y^2}+\frac{2\nu}{U_e^2}\left(\frac{\partial u}{\partial y}\right)^2 \quad [3.270]$$

It was noted that:

$$\Xi(\eta) = \frac{T - T_w}{T_e - T_w}\,2c_p\frac{T_e - T_w}{U_e^2} = r\theta \qquad [3.271]$$

The energy equation is then transformed in the new system of variables and functions:

$$\left[\sqrt{\frac{U_e}{\nu x}}\left(\frac{\partial\sqrt{\nu U_e x}\,f}{\partial\eta}\right)\right]\left[\frac{\partial\Xi}{\partial\xi} - \frac{\eta}{2x}\frac{\partial\Xi}{\partial\eta}\right] \qquad [3.272]$$

$$-\left[\frac{1}{2}\sqrt{\frac{\nu U_e}{x}}f + \sqrt{\nu U_e x}\frac{\partial f}{\partial\xi} - \frac{\eta}{2x}\frac{\partial\sqrt{\nu U_e x}\,f}{\partial\eta}\right]\left[\sqrt{\frac{U_e}{\nu x}}\frac{\partial\Xi}{\partial\eta}\right] \qquad [3.273]$$

$$= a\sqrt{\frac{U_e}{\nu x}}\frac{\partial}{\partial\eta}\sqrt{\frac{U_e}{\nu x}}\frac{\partial\Xi}{\partial\eta} + \frac{2\nu}{U_e^2}\left[\sqrt{\frac{U_e}{\nu x}}\frac{\partial}{\partial\eta}\sqrt{\frac{U_e}{\nu x}}\frac{\partial\sqrt{U_e\nu x}\,\psi}{\partial\eta}\right]^2 \qquad [3.274]$$

The additional term is rewritten as:

$$\frac{2\nu}{U_e^2}\left[\sqrt{\frac{U_e}{\nu x}}\frac{\partial}{\partial\eta}\sqrt{\frac{U_e}{\nu x}}\frac{\partial\sqrt{U_e\nu x}\,f}{\partial\eta}\right]^2 = \frac{2\nu U_e^2 U_e}{U_e^2\,\xi\nu}f_{\eta\eta}^2 = \frac{2U_e}{\xi}f_{\eta\eta}^2 \qquad [3.275]$$

which will turn into:

$$U_e f_\eta\left[\Xi_\xi - \frac{\eta}{2\xi}\Xi_\eta\right] - \left[\frac{U_e}{2\xi}f\Xi_\eta + U_e f_\xi\Xi_\eta - U_e\frac{\eta}{2\xi}f_\eta\,\theta_\eta\right] = U_e\frac{a}{\nu\xi}\Xi_{\eta\eta} + \frac{2U_e}{\xi}f_{\eta\eta}^2 \qquad [3.276]$$

Assuming the similarity hypothesis for Ξ, so $\Xi_\xi = 0$, we find that:

$$-\frac{\eta f_\eta\Xi_\eta}{2\xi} - \frac{f\Xi_\eta}{2\xi} + \frac{\eta f_\eta\Xi_\eta}{2\xi} = \frac{1}{\Pr\xi}\Xi_{\eta\eta} + \frac{2f_{\eta\eta}^2}{\xi} \qquad [3.277]$$

$$-\frac{f\Xi_\eta}{2\xi} = \frac{1}{\Pr \xi}\Xi_{\eta\eta} + \frac{2f_{\eta\eta}^2}{\xi} \quad [3.278]$$

The modified Pohlhausen equation thus becomes:

$$-\frac{f\Xi_\eta}{2\xi} = \frac{1}{\Pr \xi}\Xi_{\eta\eta} + \frac{2f_{\eta\eta}^2}{\xi} \quad [3.279]$$

$$2\Xi_{\eta\eta} + \Pr f\Xi_\eta + 4\Pr f_{\eta\eta}^2 = 0 \quad [3.280]$$

$$2\Xi'' + \Pr f\Xi' + 4\Pr f''^2 = 0 \quad [3.281]$$

with the following boundary conditions:

$$\Xi(0) = 0 \;\; ; \quad \Xi(\infty) \to \frac{2c_p\left(T_e - T_w\right)}{U_e^2} = r \quad [3.282]$$

Moreover, the parietal heat flow is zero (adiabatic wall), therefore:

$$\lambda \left.\frac{\partial T}{\partial y}\right|_{y=0} = 0 \quad [3.283]$$

and as a result:

$$\left.\frac{\partial \Xi}{\partial \eta}\right|_{y=0} = \sqrt{\frac{\nu x}{U_e}} \left.\frac{\partial r\theta}{\partial y}\right|_{y=0} = 0 \quad [3.284]$$

So $\Xi'(0) = 0$ [3.285]

3.3) If the parietal flow is not zero, we must return to the previous lines.

Let us recall that:

$$\Xi(\eta) = r\theta = r\frac{T_W - T}{T_W - T_e} \quad [3.286]$$

and therefore:

$$T_W - T = \frac{T_W - T_e}{r} \Xi \qquad [3.287]$$

Moreover, the parietal heat flow is written in absolute value, since it is clear that:

$$\frac{\partial T}{\partial y} < 0 \qquad [3.288]$$

$$\varphi_W = -\lambda \left.\frac{\partial T}{\partial y}\right|_{y=0} = \lambda \frac{T_W - T_e}{r} \left.\frac{\partial \Xi}{\partial y}\right|_{y=0} = \lambda \frac{T_W - T_e}{r} \Xi'(0) \qquad [3.289]$$

Therefore:

$$\Xi'(0) = \frac{r}{\lambda (T_W - T_e)} \varphi_W \qquad [3.290]$$

We note that $\Xi'(0)$ is, in fact, a Nusselt number.

4) Let us integrate the Polhausen equation in its more complete form:

$$2\Xi'' + \Pr f \Xi' + 4 \Pr f''^2 = 0 \qquad [3.291]$$

4.1) From the Blasius equation, we get the following:

$$f = -\frac{2f'''}{f''} \qquad [3.292]$$

$$2\Xi'' - 2\Pr \frac{f'''}{f''} \Xi' + 4 \Pr f''^2 = 0 \qquad [3.293]$$

$$\Xi'' - \Pr \frac{f'''}{f''} \Xi' = -2 \Pr f''^2 \qquad [3.294]$$

The complete solution is the sum of the general equation of the equation without the second member, plus a particular solution of the equation with the second member.

a) Equation without the second member:

$$\Xi'' - \Pr \frac{f'''}{f''} \Xi' = 0 \qquad [3.295]$$

$$\frac{\Xi''}{\Xi'} = \Pr \frac{f'''}{f''} \qquad [3.296]$$

$$Ln\,\Xi' = \Pr Ln\, f'' + Ln\, C_1 \qquad [3.297]$$

$$\Xi' = C_1 f''^{\Pr} \qquad [3.298]$$

b) For the equation with the second member, let us use the method of variation of the constant:

$$\Xi'' - \Pr \frac{f'''}{f''} \Xi' = -2 \Pr f''^2 \qquad [3.299]$$

$$\Xi' = C(\eta) f''^{\Pr} \qquad [3.300]$$

$$\Xi'' = C' f''^{\Pr} + C \Pr f''^{\Pr-1} f''' \qquad [3.301]$$

$$C' f''^{\Pr} + C \Pr f''^{\Pr-1} f''' - \Pr \frac{f'''}{f''} C f''^{\Pr} = -2 \Pr f''^2 \qquad [3.302]$$

Or, after simplification by f''^2:

$$C' f''^{\Pr} + C \Pr f''^{\Pr-1} f''' - \Pr f''' C f''^{\Pr-1} = -2 \Pr f''^2 \qquad [3.303]$$

$$C' f''^{\Pr} = -2 \Pr f''^2 \qquad [3.304]$$

$$C' = -2 \Pr f''^{2-\Pr} \qquad [3.305]$$

$$C = -\int_0^{\overline{\overline{\eta}}} 2 \Pr f''^{2-\Pr}\, d\overline{\overline{\eta}} + C_1 \qquad [3.306]$$

This shows that:

$$\Xi' = \left[-2\,\mathrm{Pr} \int_0^{\eta} \overset{=}{f}{}''^{2-\mathrm{Pr}}\, d\overset{=}{\eta} + C_1 \right] f''^{\mathrm{Pr}} \qquad [3.307]$$

By integrating:

$$\Xi = 2\,\mathrm{Pr} \int_0^{\eta} f''^{\mathrm{Pr}} \left[-\int_0^{\bar{\eta}} \overset{=}{f}{}''^{2-\mathrm{Pr}}\, d\overset{=}{\eta} + C_1 \right] d\bar{\eta} + C_2 \qquad [3.308]$$

c) Using the boundary conditions:

$$\Xi'(0) = 0 \quad ; \qquad C_1 = 0 \qquad [3.309]$$

$$\Xi(\infty) \to \frac{2c_p\left(T_e - T_w\right)}{U_e^2}\,\theta(\infty) = r \quad ; \qquad [3.310]$$

$$r = 2\,\mathrm{Pr} \int_0^{\infty} f''^{\mathrm{Pr}} \left[-\int_0^{\bar{\eta}} \overset{=}{f}{}''^{2-\mathrm{Pr}}\, d\overset{=}{\eta} \right] d\bar{\eta} + C_2 \qquad [3.311]$$

$$C_2 = 2\,\mathrm{Pr} \int_0^{\infty} f''^{\mathrm{Pr}} \left[+\int_0^{\bar{\eta}} \overset{=}{f}{}''^{2-\mathrm{Pr}}\, d\overset{=}{\eta} \right] d\bar{\eta} + r \qquad [3.312]$$

$$\Xi = 2\,\mathrm{Pr} \int_0^{\eta} f''^{\mathrm{Pr}} \left[-\int_0^{\bar{\eta}} \overset{=}{f}{}''^{2-\mathrm{Pr}}\, d\overset{=}{\eta} \right] d\bar{\eta} + 2\,\mathrm{Pr} \int_0^{\infty} f''^{\mathrm{Pr}} \left[+\int_0^{\bar{\eta}} \overset{=}{f}{}''^{2-\mathrm{Pr}}\, d\overset{=}{\eta} \right] d\bar{\eta} + r \qquad [3.313]$$

d) Ultimately:

$$\Xi = 2\,\mathrm{Pr} \int_{\eta}^{\infty} f''^{\mathrm{Pr}} \left[-\int_0^{\bar{\eta}} \overset{=}{f}{}''^{2-\mathrm{Pr}}\, d\overset{=}{\eta} \right] d\bar{\eta} + r \qquad [3.314]$$

4.2) Let us deduce the parietal heat factor r from this result, as a function of $f(\eta)$and the Prandtl number:

$$\Xi = 2\,\mathrm{Pr}\int_{\eta}^{\infty} f''^{\mathrm{Pr}}\left[-\int_0^{\bar{\bar{\eta}}} f''^{2-\mathrm{Pr}}\,d\bar{\bar{\eta}}\right]d\bar{\eta}+r \qquad [3.315]$$

$$\Xi(0) = 0 \qquad [3.316]$$

Therefore:

$$\Xi(0) = 0 = 2\,\mathrm{Pr}\int_{0}^{\infty} f''^{\mathrm{Pr}}\left[-\int_0^{\bar{\bar{\eta}}} f''^{2-\mathrm{Pr}}\,d\bar{\bar{\eta}}\right]d\bar{\eta}+r \qquad [3.317]$$

From this, we can easily deduce:

$$r = 2\,\mathrm{Pr}\int_{0}^{\infty} f''^{\mathrm{Pr}}\left[\int_0^{\bar{\bar{\eta}}} f''^{2-\mathrm{Pr}}\,d\bar{\bar{\eta}}\right]d\bar{\eta} \qquad [3.318]$$

4.3) It has been shown in the literature that a good approximation of this factor is given by:

$$r \approx \sqrt{\mathrm{Pr}} \qquad [3.319]$$

This is verified here: if $\mathrm{Pr} = 1$

Then:

$$r = 2\,\mathrm{Pr}\int_{0}^{\infty} f''\left[\int_0^{\bar{\bar{\eta}}} f''\,d\bar{\bar{\eta}}\right]d\bar{\eta} \qquad [3.320]$$

$$\int_0^{\bar{\bar{\eta}}} f''\,d\bar{\bar{\eta}} = f'(\eta) - f'(0) = f'(\eta) \qquad [3.321]$$

$$r = 2\,\mathrm{Pr}\int_{0}^{\infty} f''\,f'\,d\bar{\eta} = 2\,\mathrm{Pr}\int_{0}^{\infty} \frac{\partial}{\partial\eta}\frac{f'^2}{2}\,d\bar{\eta} = 2\,\mathrm{Pr}\left.\frac{f'^2}{2}\right|_0^{\infty} = \mathrm{Pr}(1-0) \qquad [3.322]$$

Therefore:

$$r = \Pr = 1 \qquad [3.323]$$

EXAMPLE 3.6.– Simplified study of natural convection

We want to calculate the heat flow carried by natural convection on a vertical flat wall of temperature T_w, immersed in a bath of molten sodium at temperature T_e. To do this, we will first establish the relationship between the parietal flow density φ_w and the temperature difference $T_w - T_e$, using a simplified theory.

It should be noted at this point that sodium, like any molten metal, has a very low Prandtl number.

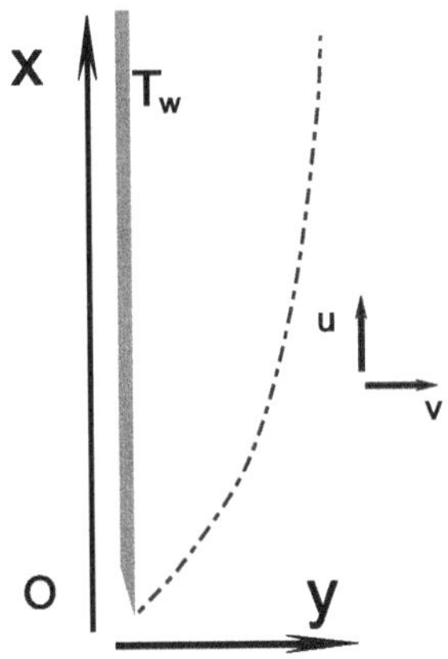

Figure 3.6. *The system of coordinates and variables. For a color version of this figure, see www.iste.co.uk/ledoux/heat3.zip*

The laws of similarity teach us that this relationship will be of the following form:

$$N_u(x) = C \Pr^m Gr_x^n \qquad [3.324]$$

where $N_u(x)$ is the local Nusselt number defined by:

$$N_u(x) = \frac{\varphi_w(x)\,x}{\lambda(T_w - T_e)} \qquad [3.325]$$

Pr is the Prandtl number of molten sodium.

$Gr(x)$ is a local Grashof number defined by:

$$Gr(x)=\frac{g\,\beta(T_w-T_e)x^3}{\nu^2} \quad [3.326]$$

β is an expansion coefficient of the fluid, such that the density of the fluid is expressed as a function of temperature by:

$$\rho(T)=\rho_e\left[1-\beta(T-T)_e\right] \quad [3.327]$$

where ρ_e is the density away from the wall (where the temperature is T_e).

NOTE.– This expression is derived from a more general definition of β:

$$\beta=\frac{1}{\rho}\frac{d\rho}{dT} \quad [3.328]$$

The main notations are summarized in a nomenclature, located at the beginning of this book.

We will start by putting the problem into an equation, then we will look for *C*, *m* and *n*, noting that for a molten metal, such as sodium, the Prandtl number is very low.

1) Writing the transfer equations

1.1) Natural convection is a boundary layer phenomenon. Use this note to write the form that continuity, momentum and energy equations will use here.

1.2) We will see two terms of the form $-\rho g$ and $-\dfrac{\partial p}{\partial x}$.

Remembering that we are in a boundary layer, and applying the fundamental principle of hydrostatics away from the wall (where the fluid is assumed to be stationary), show that this term can be put into the form:

$$\rho_e\, g\,\beta(T_w-T_e)$$

At this point, we assume that β is enough to account for the effects of variation of ρ, and in the rest of the equation, we will put:

$$\rho \approx \rho_e \tag{3.329}$$

(This simplification will make the kinematic viscosity $\nu = \dfrac{\mu}{\rho_e}$ and the thermal diffusivity $a = \dfrac{\mu c_P}{\lambda}$ appear.)

1.3) Write the boundary conditions in the case of a real fluid, then in the case of a perfect fluid.

2) Solution in the case of a molten metal

2.1) In this case, the terms containing the viscosity can be neglected in the momentum equation. Justify this approximation in a few words.

2.2) It is assumed that near the wall, the temperature of the fluid is not much different from T_w. Show that it can be assumed that u is only a function of x (locally uniform flow), and takes the following form:

$$u(x) = B\sqrt{x} \tag{3.330}$$

Give the expression for B.

2.3) We can then solve the energy equation. We will use the dimensionless variable for this:

$$\theta(x,y) = \frac{T_w - T(x,y)}{T_w - T_e} \tag{3.331}$$

a) Write the equation for $\theta(x, y)$, as well as the boundary conditions.

b) Using the change of variable:

$$\psi = \frac{2a}{B} x^{\frac{1}{2}} \tag{3.332}$$

where a is the thermal diffusivity of the fluid, find an equation whose form is familiar to you.

c) Using a fundamental result of the conduction course, give the expression of θ as a function of Ψ and y, then T as a function of x and y.

d) Calculate the parietal flow density $\varphi_w(x)$, then the Nusselt number $Nu(x)$.

e) Deduce the values of C, m and n.

SOLUTION TO EXAMPLE 3.6.–

1) Writing the transfer equations

1.1) Natural convection is a boundary layer phenomenon. Use this note to write the form that continuity, momentum and energy equations will use here.

Continuity equation:

$$\frac{\partial \rho}{\partial t}+\frac{\partial \rho u}{\partial x}+\frac{\partial \rho v}{\partial y}=0 \qquad [3.333]$$

Momentum equation:

$$\rho\left[\frac{\partial u}{\partial t}+u\frac{\partial u}{\partial x}+v\frac{\partial u}{\partial y}\right]$$

$$=-\frac{\partial p}{\partial x}+\frac{\partial}{\partial x}\left[2\mu\left(\frac{\partial u}{\partial x}\right)\right]+\frac{\partial}{\partial y}\left[\mu\left(\frac{\partial u}{\partial y}+\frac{\partial v}{\partial x}\right)\right]+\frac{\partial}{\partial x}\left[\eta\left(\frac{\partial u}{\partial x}+\frac{\partial v}{\partial y}\right)\right]+\rho F_{Vx}$$

$$\rho\frac{\partial v}{\partial t}+u\frac{\partial v}{\partial x}+v\frac{\partial v}{\partial y} \qquad [3.334]$$

$$=-\frac{\partial p}{\partial y}+\frac{\partial}{\partial x}\left[\mu\left(\frac{\partial u}{\partial y}+\frac{\partial v}{\partial x}\right)\right]+\frac{\partial}{\partial y}\left[2\mu\left(\frac{\partial u}{\partial y}\right)\right]+ +\frac{\partial}{\partial x}\left[\eta\left(\frac{\partial u}{\partial x}+\frac{\partial v}{\partial y}\right)\right]+\rho F_{Vy}$$

Energy equation:

$$\rho c_p \left(\frac{\partial T}{\partial t} + u\frac{\partial T}{\partial x} + v\frac{\partial T}{\partial y} \right) - \left(\frac{\partial p}{\partial t} + u\frac{\partial p}{\partial x} + v\frac{\partial p}{\partial y} \right)$$

$$= - \left(\frac{\partial}{\partial x}\lambda\frac{\partial T}{\partial x} + \frac{\partial}{\partial y}\lambda\frac{\partial T}{\partial y} \right) + Q + \Phi \qquad [3.335]$$

$$\Phi = \mu \left[\left(\frac{2\partial u}{\partial y} \right)\frac{\partial u}{\partial x} + \left(\frac{\partial u}{\partial y} + \frac{\partial v}{\partial x} \right)\frac{\partial u}{\partial y} + \left(\frac{\partial v}{\partial x} + \frac{\partial u}{\partial y} \right)\frac{\partial v}{\partial x} + \left(\frac{2\partial v}{\partial y} \right)\frac{\partial v}{\partial y} \right] \qquad [3.336]$$

We can simplify these equations by accepting the hypotheses:

– With the exception of density (this case will be discussed in Question 1.2), the physical properties of the liquid are constant.

– The flow is stationary.

– The boundary layer approximations apply.

– The velocities are low, there is no kinetic heating. The dissipation function Φ will be very small.

– The volume forces are reduced to gravity, which applies according to Ox:

$$\rho F_{Vx} = -\rho g \qquad [3.337]$$

It then comes to:

$$\frac{\partial \rho u}{\partial x} + \frac{\partial \rho v}{\partial y} = 0 \qquad [3.338]$$

$$\rho \left[u\frac{\partial u}{\partial x} + v\frac{\partial u}{\partial y} \right] = -\frac{\partial p}{\partial x} + \mu\frac{\partial u}{\partial y} - \rho g \qquad [3.339]$$

$$\rho c_p \left(u\frac{\partial T}{\partial x} + v\frac{\partial T}{\partial y} \right) - \left(u\frac{\partial p}{\partial x} \right) = = \lambda\frac{\partial^2 T}{\partial y^2} \qquad [3.340]$$

The dissipation function is negligible: Φ ; 0

1.2) We will see two terms of the form $-\rho g$ and $-\frac{\partial p}{\partial x}$.

Remembering that we are in a boundary layer, and applying the fundamental principle of hydrostatics away from the wall (where the fluid is assumed to be stationary), show that this term can be put into the form:

$$\rho_e g \beta(T_w - T_e) \quad [3.341]$$

At this point, we assume that β is enough to account for the effects of variation of ρ, and in the rest of the equation, we will put:

$$\rho \approx \rho_e \quad [3.342]$$

(This simplification will make the kinematic viscosity c and the thermal diffusivity a appear.)

We therefore seek to assess the term: $-\frac{\partial p}{\partial x} - \rho g$.

In this kind of question, it is necessary to consider the generating pressure (see [LED 17a]):

$$p_G = p + \rho g Z \quad [3.343]$$

where Z is a dimension taken on a vertical axis, oriented from bottom to top (in other words, in the direction and in the opposite direction of the forces of gravity). Here, the Z axis is merged with the x.

As a result:

$$p_G = p + \rho g x \quad [3.344]$$

and:

$$-\frac{\partial p}{\partial x} - \rho g = -\frac{\partial(p + \rho g x)}{\partial x} = -\frac{\partial p_G}{\partial x} \quad [3.345]$$

The generating pressure does not vary along any axis that is perpendicular to the current lines of a parallel flow. The boundary layer is an almost parallel flow. So, here: $\dfrac{\partial p_G}{\partial y} = 0$.

Outside the boundary layer, the fundamental theorem of hydrostatics applies. Indeed, in this zone, the density has a constant value ρ_e.

$$\frac{\partial\left(p_e(x) + \rho_e g x\right)}{\partial x} = 0 \qquad [3.346]$$

That is:

$$\frac{\partial p_e(x)}{\partial x} = -\frac{\partial \rho_e g x}{\partial x} = -\rho_e g \qquad [3.347]$$

$$\frac{\partial p_G}{\partial y} = 0 \qquad [3.348]$$

This implies:

$$\frac{\partial(p + \rho g x)}{\partial y} = \frac{\partial p}{\partial y} = 0 \qquad [3.349]$$

The static pressure is therefore constant on any plane that is normal to the plate. Therefore, at an abscissa x: $p(x) = p_e(x)$.

From this, we deduce:

$$-\frac{\partial p}{\partial x} = -\frac{\partial p_e}{\partial x} = \rho_e g \qquad [3.350]$$

and the studied term $-\dfrac{\partial p}{\partial x} - \rho g$ ultimately becomes:

$$-\frac{\partial p}{\partial x} - \rho g = \rho_e g - \rho g = (\rho_e - \rho) g \qquad [3.351]$$

Introducing the expansion coefficient:

$$\frac{\rho(T)}{\rho_e} = \left[1 - \beta(T - T_e)\right] \qquad [3.352]$$

$$-\frac{\partial p}{\partial x} - \rho g = \rho_e \left(1 - \frac{\rho}{\rho_e}\right) g = \rho_e \left(1 - \left[1 - \beta(T - T_e)\right]\right) g = \rho_e \ \beta(T - T_e) g$$

Furthermore, since the corresponding changes in density are relatively small, we will assume $\rho \approx \rho_e$ in the rest of the equations.

So, ultimately, the complete system is:

$$\frac{\partial u}{\partial x} + \frac{\partial v}{\partial y} = 0 \qquad [3.353]$$

$$\rho\left(u\frac{\partial u}{\partial x} + v\frac{\partial u}{\partial y}\right) = \mu\frac{\partial u}{\partial y} + \rho \ \beta(T - T_e) g \qquad [3.354]$$

$$\rho c_P\left(u\frac{\partial T}{\partial x} + v\frac{\partial T}{\partial y}\right) = \lambda\frac{\partial^2 T}{\partial y^2} \qquad [3.355]$$

1.3) For a real fluid, the boundary conditions are written as:

$$y = 0 \ ; \ u = v = 0 \ ; T = T_w \qquad [3.356]$$

$$y \to \infty \ ; \ u \to U_e \ ; T \to T_e \qquad [3.357]$$

For a perfect fluid, the parietal friction becomes zero, and we get the following:

$$y = 0 \ ; \ \frac{\partial u}{\partial y} = \ \ ; T = T_w \qquad [3.358]$$

$$y \to \infty \ ; \ u \to U_e \ ; T \to T_e \qquad [3.359]$$

2) Solution in the case of molten metal

2.1) Viscosity value

The Prandtl number $\Pr = \dfrac{\nu}{a}$ is very small. The viscosity is therefore low:

$$\nu = a\Pr \approx \varepsilon \qquad [3.360]$$

We will thus justify neglecting the viscosity terms in the equations.

2.2) The transfer equations then take the following form:

$$\frac{\partial u}{\partial x}+\frac{\partial v}{\partial y}=0 \qquad [3.361]$$

$$\rho_e\left(u\frac{\partial u}{\partial x}+v\frac{\partial u}{\partial y}\right) = \rho_e\ \beta\left(T-T_e\right)g \qquad [3.362]$$

$$\rho_e c_P\left(u\frac{\partial T}{\partial x}+v\frac{\partial T}{\partial y}\right)=\lambda\frac{\partial^2 T}{\partial y^2} \qquad [3.363]$$

It is assumed that near the wall, the temperature of the fluid is not much different from T_w. The momentum equation then becomes:

$$u\frac{\partial u}{\partial x}+v\frac{\partial u}{\partial y} = \beta\left(T_W-T_e\right)g \qquad [3.364]$$

The term on the right is constant and the equation will then assume a solution where $\dfrac{\partial u}{\partial y} = 0$, in other words, a locally uniform flow. This also makes the component v insignificant.

The equation is simplified to:

$$u\frac{\partial u}{\partial x} = \beta\left(T_W-T_e\right)g = Const. \qquad [3.365]$$

So:

$$\frac{\partial u^2}{\partial x} = 2\beta\left(T_W-T_e\right)g \qquad [3.366]$$

which becomes:

$$u^2 = 2\ \beta\left(T_W - T_e\right) g\ x + C \qquad [3.367]$$

u is zero at the origin, the constant C cancels. And:

$$u(x) = B\sqrt{x} \qquad [3.368]$$

With:

$$B = \sqrt{2\ \beta\left(T_W - T_e\right) g} \qquad [3.369]$$

2.3) We can then solve the energy equation. We define:

$$\theta(x, y) = \frac{T_w - T(x, y)}{T_w - T_e} \qquad [3.370]$$

a) $\theta(x, y)$ answers with the previous simplifications:

$$B\sqrt{x}\frac{\partial \theta}{\partial x} = a\frac{\partial^2 \theta}{\partial y^2} \qquad [3.371]$$

with the boundary conditions:

$$y = 0 \quad ; \quad \theta = 0 \qquad [3.372]$$

$$y \to \infty \quad ; \quad \theta \to 1 \qquad [3.373]$$

b) Let us change the variable:

$$\psi = \frac{2a}{B} x^{\frac{1}{2}} \qquad [3.374]$$

Then, passing from the (x, y) system to the (ψ, y) system, the change of variables gives:

$$\frac{\partial}{\partial x} = \frac{\partial \psi}{\partial x}\frac{\partial}{\partial \psi} + \frac{\partial y}{\partial x}\frac{\partial}{\partial y} = \frac{\partial \psi}{\partial x}\frac{\partial}{\partial \psi} = \frac{a}{B\sqrt{x}}\frac{\partial}{\partial \psi} \qquad [3.375]$$

$$\frac{\partial}{\partial y} = \frac{\partial \psi}{\partial y}\frac{\partial}{\partial \psi} + \frac{\partial y}{\partial y}\frac{\partial}{\partial y} = \frac{\partial}{\partial y} \qquad [3.376]$$

The equation becomes:

$$B\sqrt{x}\frac{a}{B\sqrt{x}}\frac{\partial T}{\partial \psi} = a\frac{\partial^2 T}{\partial y^2} \qquad [3.377]$$

That is:

$$\frac{\partial \theta}{\partial \psi} = \frac{\partial^2 \theta}{\partial y^2} \qquad [3.378]$$

This equation is formally identical to an unsteady heat conduction equation, where the diffusivity is equal to the unit.

c) The solution will be (see Volume 1 on conduction):

$$\theta = erf\,\frac{y}{\sqrt{4\psi}} \qquad [3.379]$$

which satisfies the boundary conditions.

d) The parietal flow $\varphi_w(x)$ will be obtained by:

$$\varphi_w(x) = \lambda\left.\frac{\partial T}{\partial y}\right|_{y=0} = \lambda(T_e - T_W)\left.\frac{\partial \theta}{\partial y}\right|_{y=0} \qquad [3.380]$$

$$\lambda(T_e - T_W)\left.\frac{\partial \theta}{\partial y}\right|_{y=0} = \lambda(T_e - T_W)\left.\frac{\partial\, erf\,\dfrac{y}{\sqrt{4\psi}}}{\partial y}\right|_{y=0} \qquad [3.381]$$

$$\lambda(T_e - T_W)\left.\frac{\partial \theta}{\partial y}\right|_{y=0} = \lambda(T_e - T_W)\frac{2}{\pi}\left.\left(\exp - \frac{y^2}{4\psi}\right)\frac{1}{\sqrt{4\psi}}\right|_{y=0} \qquad [3.382]$$

Because: $\dfrac{d\,erf(\xi)}{d\,\xi} = \dfrac{2}{\pi}\exp(-\xi^2)$

$$\psi = \frac{2a}{B}x^{\frac{1}{2}}$$

By replacing:

$$\varphi_W(x) = \frac{2}{\pi}\lambda(T_e - T_W)\frac{1}{\sqrt{4\frac{2a}{B}x^{\frac{1}{2}}}} = \frac{2}{\pi}\lambda(T_e - T_W)\sqrt{\frac{B}{8a}}x^{-0.25} \qquad [3.383]$$

$$\varphi_W(x) = \frac{2}{\pi}\lambda(T_e - T_W)$$

$$\frac{1}{\sqrt{4\frac{2a}{B}x^{\frac{1}{2}}}} = \frac{2}{\pi}\lambda(T_e - T_W)\frac{\left(2\,\beta(T_W - T_e)g\right)^{0.25}}{\sqrt{8a}}x^{-0.25} \qquad [3.384]$$

Or else:

$$\varphi_W(x) == 0.27\,\lambda(T_e - T_W)\frac{\left(\rho_e\;\beta(T_W - T_e)g\right)^{0.25}}{\left(\nu^2\right)^{0.25}}x^{-0.25}\,\mathrm{Pr}^{0.5} \qquad [3.385]$$

We used the relation: $a = \frac{\nu}{\mathrm{Pr}}$.

The Nusselt number $Nu(x)$ comes to:

$$Nu(x) = \frac{\varphi_W x}{\lambda(T_e - T_W)} = 0.42\frac{\left(\rho_e\;\beta(T_W - T_e)g\right)^{0.25}}{\left(\nu^2\right)^{0.25}}x^{0.75}\,\mathrm{Pr}^{0.5} \qquad [3.386]$$

$$Nu(x) = \frac{\varphi_W x}{\lambda(T_e - T_W)} = 0.42\frac{\left(\rho_e\;\beta(T_W - T_e)x^3\;g\right)^{0.25}}{\left(\nu^2\right)^{0.25}}\mathrm{Pr}^{0.5} \qquad [3.387]$$

The Grashof number appears:

$$Gr(x) = \frac{g\,\beta\,x^3(T_W - T_e)}{\nu^2}. \qquad [3.388]$$

$$Nu(x) = 0.27\left(Gr_x^{0.25}\right)\mathrm{Pr}^{0.5} \qquad [3.389]$$

e) This expression has the form:

$$Nu(x) = C\, Gr_x^n\, \Pr^m \qquad [3.390]$$

with $C = 0.28$

$n = 0.25$

$m = 0.5$

NOTE.– One author, Lefebvre, has shown that the exact value of C is 0.6. The difference with the result found is essentially due to the nature of the simplifying assumptions. The adoption of a uniform velocity profile is a significant factor.

EXAMPLE 3.7. Laminar film condensation on a wall

Note.– This problem could have been placed in a chapter devoted to two-phase flows. We have placed it here for two reasons:

a) Although condensation transfer is part of two-phase transfer, Chapter 1 of Volume 4, motivated by the importance of these transfers in nuclear technology, focuses on transfers with evaporation.

b) This problem remains on the mind of those that accompany it in this section, directed towards thoughts on the analytical treatment of boundary layers.

Let us look at a vapor at a saturation temperature T_{sat} corresponding to the ambient pressure p_{sat}, which is constant in all the vapor.

It is recalled that in this case, the steam condenses as soon as its temperature is lowered below T_{sat}, and that the condensation of one kilogram of steam releases a quantity of heat L_v (latent heat of vaporization).

A vertical plate maintained at a temperature T_w below T_{sat} is put in this vapor atmosphere. Condensation will occur along the plate, which will take the form of a film in stationary laminar flow from top to bottom, of increasing thickness $\delta(x)$ along the abscissa x (see figure for notations).

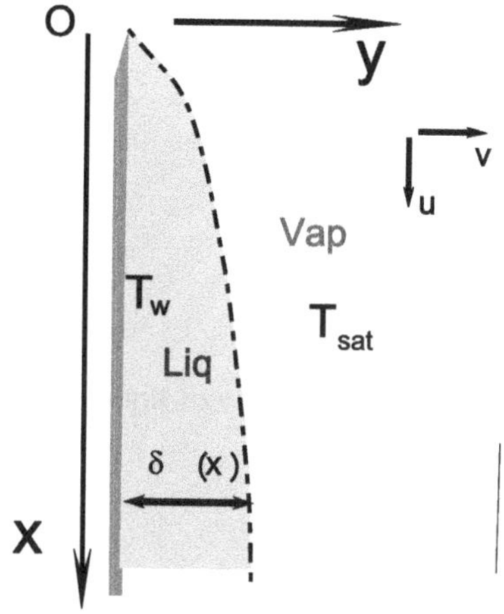

Figure 3.7. *Film of liquid condensation.* ***Liq*** *= liquid film;* ***Vap*** *= vapor atmosphere at temperature T_{sat}. For a color version of this figure, see www.iste.co.uk/ledoux/heat3.zip*

The temperature of the liquid/vapor interface is assumed to be T_{sat}. Moreover, we will assume that the variation of $\delta(x)$ is slow enough so that the flow can be considered as parallel in the transfer equations.

ρ is the density of the liquid, μ its dynamic viscosity, λ its thermal conductivity and c_p its mass heat capacity.

These physical properties are constant with the temperature.

1) Flow calculation

1.1) Show that the static pressure is constant throughout the flow.

1.2) Write the equation of the dynamics in the liquid film for the assumptions above. Solve it by writing:

– the no-slip condition at the wall;

– that the vapor is considered a perfect fluid.

1.3) Show that the longitudinal velocity profile $u(x)$ on a section of the film ($x = Const.$) can be put in the following form:

$$u(x) = K(x)\left[\frac{y}{\delta} - \frac{1}{2}\left(\frac{y}{\delta}\right)^2\right] \qquad [3.391]$$

where $K(x)$ is a function that we will give.

1.4) Find the expression for the mass flow of liquid $G(x)$ in a section of the film over a unit width.

2) Calculation of the thermal profile

We will call the convection coefficient $\alpha(x)$, which is still unknown, at the wall:

$$\alpha(x) = \frac{\varphi_w(x)}{T_{sat} - T_w} \qquad [3.392]$$

where $\varphi_w(x)$ is the parietal heat flow density.

2.1) Write the particularly simple form that the energy equation takes.

2.2) Solve this equation using the boundary conditions of the film.

2.3) Deduce the expression of $\alpha(x)$ as a function of λ and $\delta(x)$.

3) Calculation of α

3.1) The growth of the film is due to the condensation of vapor at the interface = liquid–vapor. We will assume that all the heat released by this condensation between x and $x + dx$ is transmitted by the film to the wall, between the same x and $x + dx$.

By performing heat and mass balances in a slice of film of thickness dx and using 1.4, establish the relations between:

$$\frac{dG}{dx} \text{ and } \frac{d\delta}{dx} \qquad [3.393]$$

$\dfrac{dG}{dx}$ and the thermal parameters of the system, and in particular L_v.

3.2) Deduce $\dfrac{d\delta}{dx}$, then $\delta(x)$.

3.3) Deduce the expression of $\alpha(x)$ from 2.3.

3.4) Put the result in the form of a relation between the Nusselt number:

$$Nu(x) = \frac{\alpha(x)x}{\lambda} \qquad [3.394]$$

as well as the following numbers:

Prandtl number Pr of the liquid;

Galileo number $Ga = \dfrac{g\,x^3}{\nu^2}$

where ν is the kinematic viscosity of the liquid;

Jakob number $Ja = \dfrac{L_v}{c_p\left(T_{sat} - T_w\right)}$.

SOLUTION TO EXAMPLE 3.7.–

1) Flow calculation

1.1) In the external flow, the static pressure can be considered as constant, given that we are in a light gas.

The flow in the film is a boundary layer flow. It is therefore practically parallel, and the generating pressure p_G is normally constant to its current lines, which in the adopted coordinate system will be solved by:

$$\frac{\partial p_G}{\partial y} = 0 \qquad [3.395]$$

In the adopted geometry, the strictly vertical flow p_G is defined by: $p_G = p - \rho g x$. This is done by taking the origin of the dimensions at the origin of the coordinates (which has no impact on the following, since we will derive), and noting that the axis is directed from top to bottom.

Therefore:

$$\frac{\partial p_G}{\partial y} = \frac{\partial p + \rho g x}{\partial y} = \frac{\partial p}{\partial y} = 0 \qquad [3.396]$$

Because it is clear that

$$\frac{\partial \rho g x}{\partial y} = 0 . \qquad [3.397]$$

Thus, on a horizontal plane, the pressure is constant in the film and equal to the pressure in the gas, where the pressure is uniform. The pressure is therefore constant in the whole film.

It will therefore be noted that, in the film, the gradient of generating pressure in the direction of the flow is constant; noting that the flow, *Ox*, is descending (hence the $-\ \rho g x$):

$$\frac{d\, p_G}{d\, x} = \frac{d\, p - \rho g x}{d\, x} = -\rho g \qquad [3.398]$$

1.2) The dynamics equation will be that of the laminar boundary layer, taking into account the generating pressure gradient:

$$\frac{\partial u}{\partial x} + \frac{\partial v}{\partial y} = 0 \qquad [3.399]$$

$$u\frac{\partial u}{\partial x} + v\frac{\partial v}{\partial y} = -\frac{1}{\rho}\frac{d\, p_G}{d\, x} + \nu\frac{\partial^2 u}{\partial y^2} \qquad [3.400]$$

Note that both pressure and viscosity forces are considered here through the term

$$p_G = p + \rho g Z . \qquad [3.401]$$

So, here:

$$u\frac{\partial u}{\partial x}+v\frac{\partial v}{\partial y} = g+\nu\frac{\partial^2 u}{\partial y^2} \qquad [3.402]$$

Considering the flow as practically parallel, we will get $v \approx 0$, and:

$$u\frac{\partial u}{\partial x} = \nu\frac{\partial^2 u}{\partial y^2} \qquad [3.403]$$

Moreover, the continuity equation tells us that:

$$\frac{\partial u}{\partial x} = -\frac{\partial v}{\partial y} \approx 0 \qquad [3.404]$$

The equation to solve will be reduced to:

$$\frac{g}{\nu}+\frac{\partial^2 u}{\partial y^2}=0 \qquad [3.405]$$

Equation with boundary conditions:

No-slip condition at the wall:

$$y=0 \quad ; \quad u=0 \qquad [3.406]$$

Perfect slip condition at the film boundary:

$$y=\delta \quad ; \quad \frac{\partial u}{\partial y}=0 \qquad [3.407]$$

Here, the condition is not asymptotic, given that the liquid domain is limited and limited by the thickness $\delta(x)$.

NOTE.– The equation written here neglects inertial effects in the momentum equation. It implies a balance between gravity forces, which make the film go down, as well as the viscosity forces, which hold it at the wall.

1.3) Calculation of the velocity profile in the film

Let us solve the differential equation, which is very classical:

$$\frac{\partial^2 u}{\partial y^2} = -\frac{g}{\nu} \qquad [3.408]$$

$$\frac{\partial u}{\partial y} = -\frac{g}{\nu} y + C_1 \qquad [3.409]$$

$$u = -\frac{g}{\nu}\frac{y^2}{2} + C_1\, y + C_2 \qquad [3.410]$$

The boundary conditions give us:

No-slip condition at the wall:

$$y = 0 \quad ; \quad -\frac{g}{\nu} 0 + C_1\, 0 + C_2 = 0 \quad ; \quad C_2 = 0 \qquad [3.411]$$

Perfect slip condition at the film boundary:

$$y = \delta \quad ; \quad -\frac{g}{\nu}\frac{\delta^2}{2} + C_1\, \delta = 0 \quad ; \quad C_1 = \frac{\rho g \delta}{2\nu} \qquad [3.412]$$

So:

$$u = -\frac{g}{\nu}\frac{y^2}{2} + \frac{\rho g \delta}{\nu} y \qquad [3.413]$$

Introducing the reduced variable $\frac{y}{\delta}$, the expression of the velocity becomes:

$$u = \frac{g\delta^2}{\nu}\left[-\frac{1}{2}\left(\frac{y}{\delta}\right)^2 + \frac{y}{\delta}\right] \qquad [3.414]$$

which is indeed the form:

$$u(x) = K(x)\left[\frac{y}{\delta} - \frac{1}{2}\left(\frac{y}{\delta}\right)^2\right] \qquad [3.415]$$

with:

$$K(x)=\frac{g\delta^2(x)}{\nu} \qquad [3.416]$$

The film thickness $\delta(x)$ dependent on the abscissa is to be determined by a thermal balance.

1.4) Calculation of the mass flow per unit of film width.

The mass flow rate $G(x)$ in a unit width of the film will be given by the integral:

$$G(x)=\int_0^{\delta(x)}\rho u\,dy \qquad [3.417]$$

$$G(x)=\int_0^{\delta(x)}\rho\frac{g\delta^2}{\nu}\left[-\frac{1}{2}\left(\frac{y}{\delta}\right)^2+\frac{y}{\delta}\right]dy=\int_0^{\delta(x)}\frac{g\delta}{\nu}\left[-\frac{1}{2}\left(\frac{y}{\delta}\right)^2+\frac{y}{\delta}\right]dy \qquad [3.418]$$

Using the following as variable of integration:

$$G(x)=\int_0^1\frac{\rho g\delta^2}{\nu}\left[-\frac{1}{2}\left(\frac{y}{\delta}\right)^2+\frac{y}{\delta}\right]\delta d\left(\frac{y}{\delta}\right)=\int_0^1\frac{\rho g\delta^3}{\nu}\left[-\frac{1}{2}\left(\frac{y}{\delta}\right)^2+\frac{y}{\delta}\right]d\left(\frac{y}{\delta}\right) \qquad [3.419]$$

So:

$$G(x)=\int_0^1\frac{\rho g\delta^3}{\nu}\left[-\frac{1}{2}(\xi)^2+\xi\right]d(\xi)=\frac{\rho g\delta^3}{\nu}\left(-\frac{\xi^3}{6}+\frac{\xi^2}{2}\right)_0^1 \qquad [3.420]$$

$$G(x)=\frac{1}{3}\frac{\rho g\delta^3}{\nu} \qquad [3.421]$$

2) Calculation of the thermal profile

Let $\alpha(x)$ be the convection coefficient at the wall:

$$\alpha(x)=\frac{\varphi_w(x)}{T_{sat}-T_w} \quad [3.422]$$

where $\varphi_w(x)$ is the parietal heat flow density.

2.1) We start from the energy equation for a chemically inert boundary layer:

$$u\frac{\partial T}{\partial x}+v\frac{\partial T}{\partial y} = a\frac{\partial^2 T}{\partial y^2} \quad [3.423]$$

Considering the film as a near-parallel flow, the temperature field will only be variable when it is normal to the plate. The energy equation then becomes:

$$\frac{\partial^2 T}{\partial y^2}=0 \quad [3.424]$$

2.2) The solution of this equation will be simple. The zero second derivative shows us a linear profile.

$T(y)$ having a zero second derivative will have the form:

$$T(y)=C_1 y+C_2$$

The boundary conditions for the film will be:

At the wall:

$$y=0 \quad ; \quad T(y)=C_1 0+C_2=T_W \quad ; \quad C_2=T_W \quad [3.425]$$

At the film border:

$$y=\delta \quad ; \quad T=C_1\delta+T_W=T_{SAT} \quad ; \quad C_1=\frac{T_{SAT}-T_W}{\delta} \quad [3.426]$$

Therefore:

$$T(y)=(T_{SAT}-T_W)\frac{y}{\delta}+T_W \quad [3.427]$$

2.3) The parietal flow density is calculated (in absolute value, taking into account that $\frac{\partial T}{\partial y} > 0$) by:

$$\varphi_W = \lambda \left.\frac{\partial T}{\partial y}\right|_{y=0} \qquad [3.428]$$

The temperature profile being linear; on any normal to the plate, $\frac{\partial T}{\partial y}$ is "equal to":

$$\frac{\partial T}{\partial y} = \frac{T_{SAT} - T_W}{\delta} \qquad [3.429]$$

and:

$$\varphi_W = \lambda \left.\frac{\partial T}{\partial y}\right|_{y=0} = \lambda \frac{T_{SAT} - T_W}{\delta} \qquad [3.430]$$

The convection coefficient then becomes:

$$\alpha = \frac{\varphi_W}{T_{SAT} - T_W} = \frac{\lambda}{\delta} \qquad [3.431]$$

3) Calculation of α

3.1) We establish a balance on a unit width of the wall. Let us consider a slice of thickness *dx*, located at the abscissa *x*. Condensation occurs on the upper part of the slice, which has an area *dx*. The variation of the film thickness $d\delta$ is indeed very low.

It is possible to calculate an increase in liquid mass flow $dG(x)$ between the entry and exit sides of the slice, due to condensation.

This condensation will also lead to the release of a quantity of heat that will be transferred to the wall (this will be the flow density φ_W).

We will thus have to write a mass balance and a thermal balance between the two faces of the slice.

Mass balance; the variation of the mass flow per unit width is related to the variation of the film thickness by:

$$G(x)=\frac{1}{3}\frac{\rho g\delta^3}{\nu} \qquad [3.432]$$

Therefore:

$$d\,G(x)=\frac{3}{3}\frac{\rho g\delta^2}{\nu}d\,\delta=\frac{\rho g\delta^2}{\nu}d\,\delta \qquad [3.433]$$

The mass contribution per unit width corresponds to an energy release per unit width over the thickness *dx*, which is equal to the flow at the wall in the slice $\varphi_W\, dx$ (φ_W is the flow per unit area and *dx* is the area from the slice to the wall). Indeed, as reported in 2.1, and the film thickness varying little with *x*, the heat transfer flow is practically normal to the wall. We deduce that:

$$L_v\, d\,G(x)=\varphi_W\, d\,x \qquad [3.434]$$

which can be rewritten as:

$$\frac{d\,G(x)}{d\,x}=\frac{\varphi_W}{L_v}=\frac{\lambda(T_{SAT}-T_W)}{\delta(x)L_v} \qquad [3.435]$$

3.2) Deduce $\frac{d\,\delta}{d\,x}$, then $\delta(x)$.

Combining the two expressions of $\frac{dG(x)}{dx}$ obtained above gives us:

$$\frac{d\,G(x)}{d\,x}=\frac{\rho g\delta^2}{\nu}\frac{d\,\delta}{d\,x} \qquad [3.436]$$

$$\frac{d\,G(x)}{d\,x}=\frac{\lambda(T_{SAT}-T_W)}{\delta(x)L_v} \qquad [3.437]$$

$$\frac{\rho g\delta^2}{\nu}\frac{d\,\delta}{d\,x}=\frac{\lambda T_{SAT}-T_W}{\delta(x)L_v} \qquad [3.438]$$

We obtain a differential equation for $\delta(x)$:

$$\delta^3 \frac{d\delta}{dx} = \frac{\lambda\nu(T_{SAT} - T_W)}{\rho g L_v} \quad [3.439]$$

with the boundary condition originally written as:

$$x = 0 \quad ; \quad \delta = 0 \quad [3.440]$$

This is a classical solution:

$$\frac{d\delta^4}{4dx} = \frac{\lambda\nu(T_{SAT} - T_W)}{\rho g L_v} \quad [3.441]$$

$$\delta^4 = 4\frac{\lambda\nu(T_{SAT} - T_W)}{\rho g L_v} x + C_1 \quad [3.442]$$

Using the condition at the origin:

$$x = 0 \quad ; \quad 4\frac{\lambda\nu(T_{SAT} - T_W)}{\rho g L_v} 0 \; C_1 = 0 \quad ; \quad C_1 = 0 \quad [3.443]$$

and

$$\delta(x) = \left[4\frac{\lambda\nu(T_{SAT} - T_W)}{\rho g L_v} x\right]^{0.25} \quad [3.444]$$

3.3) $\alpha(x)$ can be simply deduced from this:

$$\alpha(x) = \frac{\lambda}{\delta} = \lambda\left[4\frac{\lambda\nu(T_{SAT} - T_W)}{\rho g L_T} x\right]^{-0.25} \quad [3.445]$$

3.4) Any heat transfer law can be put in dimensionless form; we introduce the Nusselt number:

$$Nu(x) = \frac{\alpha(x)x}{\lambda} \quad [3.446]$$

As well as numbers constructed from the problem data:

Prandtl number Pr of the liquid;

Galileo number $Ga = \dfrac{g\,x^3}{\nu^2}$;

Jakob number $Ja = \dfrac{L_v}{c_p\left(T_{sat} - T_w\right)}$.

We deduce the Nusselt number from this:

$$Nu(x) = \frac{\alpha(x)x}{\lambda} = \frac{\lambda x}{\lambda}\left[4\frac{\lambda\nu\left(T_{SAT}-T_W\right)}{\rho g L_v}x\right]^{-0.25} = \left[4\frac{\lambda\nu\left(T_{SAT}-T_W\right)}{\rho g x^3 L_v}\right]^{-0.25} \quad [3.447]$$

Let us make the expressions of the various numbers of Prandtl Pr, Galileo $Ga = \dfrac{g\,x^3}{\nu^2}$ and Jakob $Ja = \dfrac{L_v}{c_p\left(T_{sat} - T_w\right)}$ appear in this expression:

$$Nu(x) = \frac{\alpha(x)x}{\lambda} = \left[4\frac{\lambda\nu\left(T_{SAT}-T_W\right)}{\rho g x^3 L_v}\right]^{-0.25} = \left[\frac{\rho\nu g x^3 L_v}{4\lambda\nu^2\left(T_{SAT}-T_W\right)}\right]^{0.25} \quad [3.448]$$

Furthermore:

$$\frac{\rho\nu}{\lambda} = \frac{\mu}{\lambda} = \frac{\mu c_p}{\lambda c_p} = \frac{\Pr}{c_p} \quad [3.449]$$

$$Nu(x) = \left[\frac{\rho\nu g x^3 L_v}{4\lambda\nu^2\left(T_{SAT}-T_W\right)}\right]^{0.25} = \left[\frac{\Pr}{4}\frac{g x^3}{\nu^2}\frac{L_v}{c_P\left(T_{SAT}-T_W\right)}\right]^{0.25} \quad [3.450]$$

which can be rewritten as:

$$Nu(x) = \left[\frac{\Pr}{4}G_a\,J_a\right]^{0.25} = 0.707\,\Pr^{0.25}\,G_a^{\;0.25}\,J_a^{\;0.25} \quad [3.451]$$

NOTE.– Two new dimensionless numbers appear here. Like all dimensionless criteria, they can be interpreted as the ratio between the effects of two transport phenomena.

It is already known that the Prandtl number $\Pr = \dfrac{\nu}{a}$ compares viscosity and thermal diffusivity as two coefficients, possibly characteristic of transient phenomena.

The Galileo number $Ga = \dfrac{g\,x^3}{\nu^2}$ compares two characteristic accelerations of gravity and viscosity. It is central in a problem where the motion of the film results from a balance of these two forces. ν is the kinematic viscosity of the liquid.

The Jakob number $Ja = \dfrac{L_v}{c_p\left(T_{sat} - T_w\right)}$ compares a vaporization heat to a specific enthalpy difference, resulting from conductive transport.

3.6. External boundary layers: integral methods

3.6.1. *Principle of the integral method*

Integral methods, which are perhaps not known well enough, provide an elegant solution to quickly solve some transfer problems (including heat and mass) in laminar and turbulent boundary layers.

They allow the economy of a direct solution of the equations when the results sought are limited to the boundary layer scales, and especially to the parietal transfer.

As we will see, we need to only assume a self-similarity of the profiles; the accuracy will depend on the proximity between the assumed profile and the real profile.

Profiles approximating the one resulting from the Blasius analysis can thus be fed back into a heat transfer problem with a good success rate.

Similarly, in turbulent regimes, the so-called "power" profiles will prove particularly useful.

We will only present these methods here by restricting ourselves to dynamic boundary layers. We should not lose sight of the fact that these methods are also very effective in heat and mass transfer problems, including in the presence of parietal chemical reactions.

Incidentally, the reader might be surprised by the attention given here to laminar layers, which have more of an academic value than a practical one, at the technological level. If we can devote the time they deserve, the exercises proposed are intended to open the reader's mind to reasonings which they may not always be accustomed to. The field of the boundary layer calls extensively on the qualities of the physicist, in the understanding of mechanics and in the practice of orders of magnitude. This last point is decisive in the handling of equations, which can be impractical without reasoned simplification.

Although these methods can, to a certain extent, be applied to compressible flows (whose density is not constant), for the sake of simplicity, we will choose here to present them systematically for fluids whose properties are constant, notably with respect to temperature.

3.6.2. *Integral methods for an external boundary layer on a flat plate, in Cartesian coordinates*

The three equations of fluid mechanics, or in this case, boundary layer equations, are written as:

$$\frac{\partial u}{\partial x}+\frac{\partial v}{\partial y}=0 \qquad [3.452]$$

$$u\frac{\partial u}{\partial x}+v\frac{\partial u}{\partial y}=-\frac{1}{\rho}\frac{d\,p_G}{d\,x}+\nu\frac{\partial^2 u}{\partial y^2}=U_e\frac{d\,U_e}{d\,x}+\nu\frac{\partial^2 u}{\partial y^2}$$

$$u\frac{\partial T}{\partial x}+v\frac{\partial T}{\partial y}= a\frac{\partial^2 T}{\partial y^2} \qquad [3.453]$$

Here, we consider the possibility of a variation of the external flow speed[1]: $U_e(x)$

1 Integral methods were largely developed, by von Kármán in particular, to deal with problems of interaction between a real fluid and obstacles. In this case, the potential (external) flow is highly variable (see the reminders on this point in section 3.2).

The identity $-\dfrac{1}{\rho}\dfrac{d\,p_G}{d\,x} = U_e \dfrac{d\,U_e}{d\,x}$ [3.454]

has been used here. Let us recall (see **Chapter 3, section 3.2.1**) that we can, in fact, find the Bernoulli equation in the external flow, which is considered as perfect.

3.6.2.1. *Integral method for the momentum equation*

We are placed here in the framework of boundary layer approximations; as the flow is "very close" to a parallel flow, p_G is constant on any plane perpendicular to the current lines of the flow.

Let us integrate the continuity and momentum equations on y between 0 and infinity.

The first gives us:

$$\int_0^\infty \frac{\partial u}{\partial x} dy = -\int_0^\infty \frac{\partial v}{\partial y} dy = -v_\infty \qquad [3.455]$$

v_∞ is not zero (the flow is "deviated" towards the exterior, because of the mass flow "deficit" in the boundary layer).

$$\int_0^\infty \left(u\frac{\partial u}{\partial x} + v\frac{\partial u}{\partial y} \right) dy = \int_0^\infty \nu \frac{\partial^2 u}{\partial y^2} dy = \nu \frac{\partial u}{\partial y}\Big|_0^\infty = -\frac{1}{\rho}\mu\frac{\partial u}{\partial y} + U_e \frac{d\,U_e}{d\,x}$$
$$= \frac{\tau_W}{\rho} + U_e \frac{d\,U_e}{d\,x} \qquad [3.456]$$

where τ_W is the parietal stress.

We define the friction coefficient C_f:

$$C_f = \frac{2\tau_W}{\rho U_e^2} \qquad [3.457]$$

$$\int_0^\infty \left(u\frac{\partial u}{\partial x} + v\frac{\partial u}{\partial y} \right) dy = \nu \frac{\partial u}{\partial y}\Big|_0^\infty = -\frac{1}{\rho}\mu\frac{\partial u}{\partial y} = \frac{\tau_W}{\rho} \qquad [3.458]$$

$$\int_0^\infty \left(u\frac{\partial u}{\partial x} + v\frac{\partial u}{\partial y} \right) dy = \int_0^\infty \left(\frac{\partial u^2}{\partial x} - u\frac{\partial u}{\partial x} + \frac{\partial u v}{\partial y} - u\frac{\partial v}{\partial y} \right) dy = \int_0^\infty \left(\frac{\partial u^2}{\partial x} + \frac{\partial u v}{\partial y} \right) dy \quad [3.459]$$

because $-u\frac{\partial u}{\partial x} - u\frac{\partial v}{\partial y} = -u\left(\frac{\partial u}{\partial x} + \frac{\partial v}{\partial y} \right) = 0$ [3.460]

due to the continuity equation.

Furthermore:

$$\int_0^\infty \frac{\partial u v}{\partial y} dy = U_e v_\infty = -U_e \int_0^\infty \frac{\partial u}{\partial x} dy \quad [3.461]$$

$$\int_0^\infty \left(u\frac{\partial u}{\partial x} + v\frac{\partial u}{\partial y} \right) dy = = \int_0^\infty \left(\frac{\partial u^2}{\partial x} + \frac{\partial u v}{\partial y} \right) dy = \int_0^\infty \left(\frac{\partial u^2}{\partial x} - U_e \frac{\partial u}{\partial x} \right) dy = \frac{\tau_W}{\rho} \quad [3.462]$$

$$\int_0^\infty \left(\frac{\partial u^2}{\partial x} - U_e \frac{\partial u}{\partial x} \right) dy = \frac{d}{dx} \int_0^\infty \left(u^2 - U_e u \right) dy = \frac{\tau_W}{\rho} \quad [3.463]$$

These expressions are scaled by dividing by U_e^2.

We then define a momentum thickness:

$$\delta_2 = \int_0^\infty \left(1 - \frac{u}{U_e} \right) \frac{u}{U_e} dy \quad [3.464]$$

and we introduce the friction coefficient: $C_f = \frac{2\tau_W}{\rho U_e^2}$ [3.465]

Finally, an expression caused by the von Kármán comes into play:

$$\frac{d\delta_2}{dx} = \frac{C_f}{2} \quad [3.466]$$

3.6.2.2. *Integral method for the energy equation*

The same treatment can be applied to the energy equation in the simple form:

$$u\frac{\partial T}{\partial x}+v\frac{\partial T}{\partial y}= a\frac{\partial^2 T}{\partial y^2} \qquad [3.467]$$

Let us recall that we have established the following from the continuity equation:

$$\int_0^\infty \frac{\partial u}{\partial x}dy=-\int_0^\infty \frac{\partial v}{\partial y}dy=-v_\infty \qquad [3.468]$$

v_∞ is not zero (the flow is "deviated" towards the exterior, because of the mass flow "deficit" in the boundary layer).

Integrating the energy equation between the plate ($y=0$) and infinity:

$$\int_0^\infty\left(u\frac{\partial T}{\partial x}+v\frac{\partial T}{\partial y}\right)dy=\int_0^\infty a\frac{\partial^2 T}{\partial y^2}dy=a\frac{\partial u}{\partial y}\bigg|_0^\infty=-\frac{1}{\rho C_P}\lambda\frac{\partial u}{\partial y}\bigg|_0=\frac{\varphi_W}{\rho C_P} \qquad [3.469]$$

where φ_W is the parietal heat flow density.

We used:

$$\int_0^\infty a\frac{\partial^2 T}{\partial y^2}dy=\frac{\lambda}{\rho c_P}\frac{\partial T}{\partial y}\bigg|_0^\infty=\frac{-\lambda}{\rho c_P}\frac{\partial T}{\partial y}\bigg|_W=\frac{\varphi_W}{\rho c_P} \qquad [3.470]$$

Let us recall the definition of the thermal Stanton number S_t:

$$St=\frac{\varphi_W}{\rho c_P U_e\left(T_W-T_e\right)} \qquad [3.471]$$

written for the case where the parietal temperature is higher.

Let us also define a reduced temperature θ:

$$\theta=\frac{T(x,y)-T_W}{T_e-T_W} \qquad [3.472]$$

θ is, by definition, dimensionless. It is equal to 0 at the wall and tends from 1 to infinity.

The following entries implicitly assume that the parietal temperature and the external flow temperature are constant:

$$\int_0^\infty \left(u\frac{\partial T}{\partial x} + v\frac{\partial T}{\partial y} \right) dy = \int_0^\infty \left(\frac{\partial uT}{\partial x} - T\frac{\partial u}{\partial x} + \frac{\partial Tv}{\partial y} - T\frac{\partial v}{\partial y} \right) dy = \int_0^\infty \left(\frac{\partial uT}{\partial x} + \frac{\partial vT}{\partial y} \right) dy$$

For $-T\dfrac{\partial u}{\partial x} - T\dfrac{\partial v}{\partial y} = -T\left(\dfrac{\partial u}{\partial x} + \dfrac{\partial v}{\partial y} \right) = 0$ because of the continuity equation.

We have therefore established that:

$$\int_0^\infty \left(u\frac{\partial T}{\partial x} + v\frac{\partial T}{\partial y} \right) dy = \int_0^\infty \left(\frac{\partial uT}{\partial x} + \frac{\partial vT}{\partial y} \right) dy \qquad [3.473]$$

$$\int_0^\infty \left(u\frac{\partial T}{\partial x} + v\frac{\partial T}{\partial y} \right) dy = \frac{\varphi_W}{\rho C_P} \qquad [3.474]$$

Thus:

$$\int_0^\infty \left(\frac{\partial uT}{\partial x} + \frac{\partial vT}{\partial y} \right) dy = \frac{\varphi_W}{\rho C_P} \qquad [3.475]$$

Furthermore, we can show that:

$$\int_0^\infty \frac{\partial vT}{\partial y} dy = v_\infty T_e = T_e \int_0^\infty \frac{\partial v}{\partial y} dy = T_e \int_0^\infty -\frac{\partial u}{\partial x} dy = \frac{\partial}{\partial x} \int_0^\infty -uT_e \, dy \qquad [3.476]$$

This leads to:

$$\int_0^\infty \left(\frac{\partial uT}{\partial x} + \frac{\partial vT}{\partial y} \right) dy = \frac{\partial}{\partial x} \int_0^\infty (uT - uT_e) dy = \frac{-\varphi_W}{\rho C_P} \qquad [3.477]$$

These expressions are scaled by dividing by $U_e\left(T_e - T_W\right)$:

$$\frac{\partial}{\partial x}\int_0^\infty \frac{\left(uT-uT_e\right)}{U_e\left(T_e-T_W\right)}dy = \frac{\partial}{\partial x}\int_0^\infty \frac{u}{U_e}\frac{\left(T-T_e\right)}{\left(T_e-T_W\right)}dy = \frac{\partial}{\partial x}\int_0^\infty \frac{u}{U_e}\frac{\left(T-T_e\right)}{\left(T_e-T_W\right)}dy$$

$$\frac{\partial}{\partial x}\int_0^\infty \frac{u}{U_e}\frac{\left(T-T_e\right)}{\left(T_e-T_W\right)}dy = \frac{\partial}{\partial x}\int_0^\infty \frac{u}{U_e}\left(\theta-1\right)dy = \frac{\varphi_W}{\rho C_P U_e\left(T_e-T_W\right)}$$

So:

$$\frac{\varphi_W}{\rho C_P U_e\left(T_W-T_e\right)} = \frac{\partial}{\partial x}\int_0^\infty \frac{u}{U_e}\left(\theta-1\right)dy \qquad [3.478]$$

On the left, we can see the Stanton number to the nearest sign.

We then define a "thermal" thickness:

$$\delta_\theta = \int_0^\infty (1-\theta)\frac{u}{U_e}dy \qquad [3.479]$$

The integral equation is then written as:

$$St\left(x\right) = \frac{d\,\delta_\theta}{d\,x} \qquad [3.480]$$

A right "*d*" has been introduced here, since δ_θ is only a function of x.

This expression is analogous, in thermal science, to that of von Kármán, in dynamics.

NOTE.– It is important not to confuse the "thermal" thickness δ_θ, integrating the reduced temperature values over the entire boundary layer height, with the thermal boundary layer thickness δ_{th}, which represents the abscissa at which the conventional value reaches $\delta_{th} = 0.99$.

3.6.2.3. *Feature of the turbulent regime*

We will apply this method to the average values of the velocity $\overline{u}\left(x,y\right)$, and the temperature $\overline{T}\left(x,y\right)$, in a turbulent regime:

$$\frac{\partial \bar{u}}{\partial x} + \frac{\partial \bar{v}}{\partial y} = 0 \quad [3.481]$$

$$\bar{u}\frac{\partial \bar{u}}{\partial x} + \bar{v}\frac{\partial \bar{u}}{\partial y} = -\frac{1}{\rho}\frac{d\,p_G}{d\,x} + \varepsilon_V \frac{\partial^2 \bar{u}}{\partial y^2} = U_e \frac{d\,U_e}{d\,x} + \varepsilon_V \frac{\partial^2 \bar{u}}{\partial y^2} \quad [3.482]$$

$$\bar{u}\frac{\partial \bar{T}}{\partial x} + \bar{v}\frac{\partial \bar{T}}{\partial y} = \varepsilon_\theta \frac{\partial^2 \bar{T}}{\partial y^2} \quad [3.483]$$

The previous calculations established in laminar flow can be formally repeated step by step. A difference could appear in the expression of the tangential stress or of the thermal flow. On the wall, only the molecular terms persist and the expressions of ε_V and ε_θ are not involved anymore. The expressions written above remain valid. In the turbulent case:

$$\int_0^\infty \left(u\frac{\partial u}{\partial x} + v\frac{\partial u}{\partial y} \right) dy = \nu \frac{\partial u}{\partial y}\Big|_0^\infty = -\frac{1}{\rho}\mu\frac{\partial u}{\partial y} = \frac{\tau_W}{\rho} \quad [3.484]$$

$$\int_0^\infty a\frac{\partial^2 T}{\partial y^2} dy = \frac{\lambda}{\rho c_P}\frac{\partial T}{\partial y}\Big|_0^\infty = \frac{-\lambda}{\rho c_P}\frac{\partial T}{\partial y}\Big|_W = \frac{\varphi_W}{\rho c_P} \quad [3.485]$$

3.6.2.4. *Summary: using the integral method*

In summary, the integral method is based on two forms of the momentum equation and the energy equation.

von Kármán equation:

$$C_f = \frac{d\,\delta_2}{d\,x} \quad [3.486]$$

Energy equation:

$$St(x) = \frac{d\,\delta_\theta}{d\,x} \quad [3.487]$$

The integral method is based on these expressions: assuming a form for the profiles of

$$\frac{u}{U_e} = f\left(\frac{y}{\delta(x)}\right) \text{ and } \theta = g\left(\frac{y}{\delta_{th}(x)}, P_r\right). \qquad [3.488]$$

These profiles can involve instantaneous values (permanent in time) for the laminar regime, or average values for the turbulent regime.

The profiles are not generally asymptotic. $\frac{u}{U_e}$ is equal to 1 for any ordinate greater than $\delta(x)$, and θ for any ordinate greater than δ_θ.

We calculate a value, more or less approximate according to the fineness of the chosen expression for f or g, of the ratio of δ_2 and δ_θ to δ or δ_{th}. The forms of f and g give access to the gradient at the wall of $\frac{u}{U_e}$ and θ, therefore making it possible to calculate the friction coefficient and the Stanton number as a function of the thicknesses. The integral equations thus lead to two differential equations met by the thicknesses δ and δ_{th}. Note that we impose a self-similarity of the flow through this expression.

The efficiency of the method comes from the fact that we integrate on the boundary layer, which minimizes the effects of an error in the profile, compared to a direct calculation of the parietal stress at the wall by local derivation at $y = 0$.

3.7. Examples of applications of integral methods

EXAMPLE 3.8. Laminar boundary layer of a Newtonian fluid.

We consider the boundary layer developed by a flat plate in the path of the wind, in a uniform non-Newtonian fluid flow. The velocity U_e of the external flow is constant. The external flow temperature is T_e, and the wall temperature is T_w.

We will assess the evolution of the boundary layer thickness $\delta(x)$, the friction and the parietal heat transfer using the integral method. Three approximations of the form of the velocity profile are proposed. We will restrict ourselves to an elementary thermal profile; the reader may wish to refine the technique with a polynomial profile.

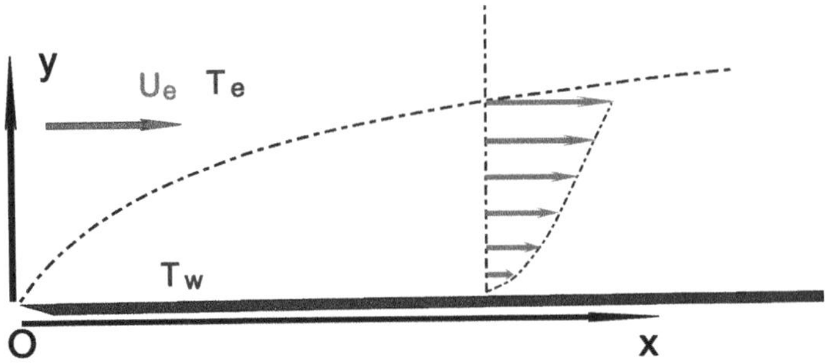

Figure 3.8. *Flat plate in the path of the wind. For a color version of this figure, see www.iste.co.uk/ledoux/heat3.zip*

1) Dynamic problem

Applying the integral method for the three profile forms assumed below, give an approximate expression for the evolution of the boundary layer thickness each time, through the $\frac{\delta(x)}{x}$ ratio and the friction coefficient $C_f(x)$.

Compare these results to those of the Blasius analysis.

Which calculations do you find most satisfactory?

1.1) The first approximation will be the most basic one imaginable:

$$\text{For } y \le \delta \quad ; \quad \frac{u}{U_e} = \frac{y}{\delta} \qquad [3.489]$$

$$\text{For } y > \delta \quad ; \quad \frac{u}{U_e} = 1 \qquad [3.490]$$

1.2) The second approximation is based on a more realistic boundary layer profile:

$$\text{For } y \le \delta \quad ; \quad \frac{u}{U_e} = \frac{3y}{2\delta} - \frac{1}{2}\left(\frac{y}{\delta}\right)^3 \qquad [3.491]$$

$$\text{For } y > \delta \quad ; \quad \frac{u}{U_e} = 1 \qquad [3.492]$$

1.3) The third approximation, which will also prove to be quite realistic, takes the following form:

For $y \leq \delta \quad ; \quad \dfrac{u}{U_e} = \sin\left(\dfrac{\pi}{2}\dfrac{y}{\delta}\right)$ [3.493]

For $y > \delta \quad ; \quad \dfrac{u}{U_e} = 1$ [3.494]

2) Thermal problem

The shape of the velocity profile here will be adopted from the second approximation of Question 1:

For $y \leq \delta \quad ; \quad \dfrac{u}{U_e} = \dfrac{3y}{2\delta} - \dfrac{1}{2}\left(\dfrac{y}{\delta}\right)^3$ [3.495]

For $y > \delta \quad ; \quad \dfrac{u}{U_e} = 1$ [3.496]

For the reduced temperature profile form, the following form will be adopted:

The reduced variable:

For $\theta = \dfrac{T - T_W}{T_e - T_W}$ [3.497]

For $y \leq \delta_{th} \quad ; \quad \theta = \dfrac{y}{\delta_{th}}$ [3.498]

For $y > \delta_{th} \quad ; \quad \theta = 1$ [3.499]

It will be assumed that the dynamic and thermal boundary layers have the same thickness: $\delta_{th} = \delta$.

It is also known that this corresponds to a Prandtl number that is equal to the unit.

Find the expression for the local Stanton number.

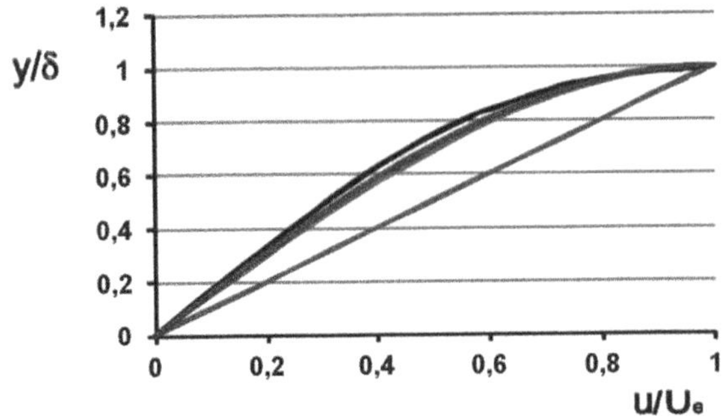

Figure 3.9. *Different approximations of the Blasius boundary layer profile. Black: exact solution (Blasius), Blue: approximation 1, linear, Red: approximation 2, polynomial, Green: approximation 3, sinusoidal. For a color version of this figure, see www.iste.co.uk/ledoux/heat3.zip*

SOLUTION TO EXAMPLE 3.8.–

1) Dynamic problem

Let us recall the von Kármán expression, which expresses the fundamental principle of dynamics in an integral form. In the case of a flow without an external velocity gradient, this equation is written as:

$$\frac{d\,\delta_2}{d\,x} = \frac{C_f}{2} \tag{3.500}$$

where:

$$\delta_2 = \int_0^{\infty}\left(1 - \frac{u}{U_e}\right)\frac{u}{U_e}\,dy \tag{3.501}$$

$$C_f = \frac{2\,\tau_w}{\rho U_e^2} \tag{3.502}$$

1.1) Linear approximation

$$\text{For } y \leq \delta \quad ; \quad \frac{u}{U_e} = \frac{y}{\delta} \tag{3.503}$$

$$\text{For } y > \delta \quad ; \quad \frac{u}{U_e} = 1 \qquad [3.504]$$

The definition of the momentum thickness δ_2 makes it possible to find its relation to the boundary layer thickness δ.

$$\delta_2 = \int_0^{\infty}\left(1-\frac{u}{U_e}\right)\frac{u}{U_e}dy = \int_0^{\delta}\left(1-\frac{y}{\delta}\right)\frac{y}{\delta}\,\delta\, d\frac{y}{\delta} \qquad [3.505]$$

Let us perform the change of variables $\xi = \dfrac{y}{\delta}$. The integration is carried out from 0 to 1, given that beyond $y = \delta$, the integrand is zero.

$$\delta_2 = \delta\int_0^1 (1-\xi)\xi\, d\xi \qquad [3.506]$$

$$\delta_2 = \delta\left(\frac{\xi^2}{2}-\frac{\xi^3}{3}\right)\Bigg|_0^1 = \frac{\delta}{6} \qquad [3.507]$$

Moreover, the slope of the velocity profile allows us to write[2] the tangential parietal stress, then the C_f:

$$\tau_W = \mu\frac{\partial u}{\partial y}\bigg|_{y=0} = \mu\frac{\partial}{\partial y}U_e\frac{u}{U_e}\bigg|_{y=0} = \frac{\mu U_e}{\delta} \qquad [3.508]$$

This is where the main source of errors in this type of calculation happens. The derivation depends on the form of the chosen profile. It should be noted in particular that from this point of view, approximations 2 and 3 are much more suitable.

In this case, we note that the tangential stress is, in fact, constant in the boundary layer:

$$C_f = \frac{2\tau_w}{\rho U_e^2} \qquad [3.509]$$

2 It is at this level and in this type of calculation that the main source of error is located.

$$C_f = \frac{2\mu U_e}{\rho U_e^2\, \delta} = \frac{2\mu}{\rho U_e \delta} \qquad [3.510]$$

The von Kármán equation is then written as:

$$\frac{1}{6}\frac{d\delta}{dx} = \frac{\mu}{\rho U_e \delta} \qquad [3.511]$$

$$\delta \frac{d\delta}{dx} = \frac{6\mu}{\rho U_e} \qquad [3.512]$$

By introducing the axial Reynolds number $R_x = \frac{U_e\, x}{\nu} = \frac{\mu U_e\, x}{\rho}$, and taking into account a boundary layer of zero thickness at the origin, the differential equation immediately gives:

$$\frac{\delta^2}{2} = \frac{6\mu x}{\rho U_e} \qquad [3.513]$$

$$\delta(x) = \sqrt{\frac{12\nu x}{U_e}} \qquad [3.514]$$

$$\frac{\delta(x)}{x} = \frac{3.46}{R_x^{1/2}} \qquad [3.515]$$

C_f then comes out to:

$$C_f = \frac{2\mu}{\rho U_e \delta} = \frac{2\nu}{U_e}\sqrt{\frac{U_e}{12\nu x}} \qquad [3.516]$$

$$\frac{C_f}{2} = \frac{\nu}{U_e}\sqrt{\frac{U_e}{12\nu x}} = \frac{0.289}{R_x^{1/2}} \qquad [3.517]$$

These results are to be compared with those of Blasius.

$$\frac{\delta(x)}{x} = \frac{4.95}{R_x^{1/2}} \quad ; \quad \frac{C_f}{2} = \frac{0.332}{R_x^{1/2}} \qquad [3.518]$$

Note that the form of $\frac{\delta(x)}{x}$ and $\frac{C_f}{2}$, to the nearest constant, is found.

This comparison shows a 30% error on δ, and a 13% error on friction C_f. It should be noted that this relatively rough approximation gives an already acceptable result on the friction, which, on a practical level, is the most important parameter.

1.2) Polynomial approximation

$$\text{For } y \leq \delta \quad ; \quad \frac{u}{U_e} = \frac{3y}{2\delta} - \frac{1}{2}\left(\frac{y}{\delta}\right)^3 \qquad [3.519]$$

$$\text{For } y > \delta \quad ; \quad \frac{u}{U_e} = 1 \qquad [3.520]$$

The stages of reasoning will be the same as before.

Let us establish the relation of momentum thickness δ_2 to boundary layer thickness δ.

$$\delta_2 = \int_0^\infty \left(1 - \frac{u}{U_e}\right)\frac{u}{U_e} dy = \int_0^\delta \left(1 - \frac{3y}{2\delta} + \frac{1}{2}\left(\frac{y}{\delta}\right)^3\right)\left[\frac{3y}{2\delta} - \frac{1}{2}\left(\frac{y}{\delta}\right)^3\right]\delta\, d\frac{y}{\delta} \qquad [3.521]$$

Let us perform the change of variables $\xi = \frac{y}{\delta}$. The integration is carried out from 0 to 1, given that beyond $y = \delta$, the integrand is zero.

$$\delta_2 = \delta \int_0^1 \left(1 - \frac{3\xi}{2} + \frac{1}{2}\xi^3\right)\left[\frac{3\xi}{2} - \frac{1}{2}\xi^3\right] d\xi \qquad [3.522]$$

By expanding the product of the parentheses of the integrand, and integrating the resulting monomials term by term, we get:

$$\delta_2 = \delta\left[-\frac{\xi^7}{28} + \frac{3\xi^5}{10} - \frac{\xi^4}{8} - \frac{9\xi^3}{12} + \frac{3\xi^2}{4}\right]_0^1 \qquad [3.523]$$

$$\delta_2 = 0.139\,\delta$$

Moreover, the slope of the velocity profile allows us to write the tangential parietal stress, then the C_f:

$$\tau_W = \mu \frac{\partial u}{\partial y}\bigg|_{y=0} = \mu \frac{\partial}{\partial y} U_e \frac{u}{U_e}\bigg|_{y=0} \quad [3.524]$$

$$= \mu \frac{\partial}{\partial y} U_e \left(\frac{3y}{2\delta} - \frac{1}{2}\left(\frac{y}{\delta}\right)^3 \right)\bigg|_{y=0} = \mu U_e \left(\frac{3y}{2\delta} - \frac{3}{2}\frac{y^2}{\delta^3} \right)\bigg|_0^1 \quad [3.525]$$

$$\tau_W = \frac{3\mu U_e}{2\delta} \quad [3.526]$$

$$\frac{C_f}{2} = \frac{\tau_W}{\rho U_e^2} = \frac{3\nu}{2\delta U_e} \quad [3.527]$$

The von Kármán equation is then written as:

$$0.139 \frac{d\delta}{dx} = \frac{3\nu}{2\delta U_e} \quad [3.528]$$

$$\frac{d\delta^2}{dx} = \frac{21.58\nu}{U_e} \quad ; \quad x = 0 \; ; \; \delta = 0 \quad [3.529]$$

By introducing the axial Reynolds number $R_x = \frac{U_e x}{\nu} = \frac{\mu U_e x}{\rho}$, the differential equation immediately gives:

$$\delta^2 = 21.58 \frac{\nu x}{U_e} \quad [3.530]$$

$$\delta(x) = \sqrt{\frac{21.58\,\nu x}{U_e}} = 4.64\sqrt{\frac{\nu x}{U_e}} \quad [3.531]$$

$$\frac{\delta(x)}{x} = \frac{4.64}{R_x^{1/2}} \quad [3.532]$$

The coefficient of friction then comes to:

$$\frac{C_f}{2} == \frac{3\nu}{2\delta U_e} = \frac{3\nu}{2*4.64\, U_e}\sqrt{\frac{U_e}{\nu x}} \quad [3.533]$$

$$\frac{C_f}{2} = \ = \frac{0.322}{R_x^{1/2}} \quad [3.534]$$

This comparison shows a 6.3% error on δ and a 2.9% error on friction C_f. We see that the form of the profile adopted, which is very close to the Blasius profile, gives an excellent approximation.

1.3) Sinusoidal approximation

$$\text{For } y \leq \delta \quad ; \quad \frac{u}{U_e} = \sin\left(\frac{\pi}{2}\frac{y}{\delta}\right) \quad [3.535]$$

$$\text{For } y > \delta \quad ; \quad \frac{u}{U_e} = 1 \quad [3.536]$$

The process remains the same as before.

Let us look for the relation between the momentum thickness δ_2 and the boundary layer thickness δ:

$$\delta_2 = \int_0^\infty \left(1 - \frac{u}{U_e}\right)\frac{u}{U_e}dy = \int_0^\delta \left(1 - \sin\left(\frac{\pi}{2}\frac{y}{\delta}\right)\right)\sin\left(\frac{\pi}{2}\frac{y}{\delta}\right) dy \quad [3.537]$$

Let us put: $\xi = \dfrac{y}{\delta}$

$$\delta_2 = \delta \int_0^1 \left(1 - \sin\left(\frac{\pi}{2}\xi\right)\right)\sin\left(\frac{\pi}{2}\xi\right) d\xi \quad [3.538]$$

The integration is conducted from 0 to 1, given that beyond $y = \delta$, the integrand is zero.

Furthermore, recalling that $\cos 2\alpha = \cos^2\alpha - \sin^2\alpha = 2\cos^2\alpha - 1$:

$$\delta_2 = \delta \int_0^1 \left(\sin\left(\frac{\pi}{2}\xi\right) - \sin^2\left(\frac{\pi}{2}\xi\right) \right) d\xi \quad [3.539]$$

$$= \delta \int_0^1 \left(\sin\left(\frac{\pi}{2}\xi\right) - \frac{1 - \cos 2\frac{\pi}{2}\xi}{2} \right) d\xi \quad [3.540]$$

$$\delta_2 = \delta \left[-\frac{2}{\pi}\cos\left(\frac{\pi}{2}\xi\right)\Bigg|_0^1 - \frac{\xi}{2}\Bigg|_0^1 + \frac{\sin \pi\xi}{2\pi}\Bigg|_0^1 \right] \quad [3.541]$$

$$= \delta\left(\frac{2}{\pi} - \frac{1}{2}\right) + 0 = \frac{4-\pi}{2\pi}\delta = 0.137\,\delta \quad [3.542]$$

$$\delta_2 = 0.137\,\delta \quad [3.543]$$

The friction is related to the boundary layer thickness by:

$$\tau_W = \mu \frac{\partial u}{\partial y}\Bigg|_{y=0} = \mu \frac{\partial}{\partial y} U_e \frac{\pi}{2} \cos\left(\frac{\pi}{2}\xi\right)\Bigg|_{y=0} = \frac{\pi}{2}\frac{\mu U_e}{\delta} \quad [3.544]$$

$$C_f = \frac{2\tau_w}{\rho U_e^2} = \pi \frac{\nu}{U_e \delta} \quad [3.545]$$

$$\frac{C_f}{2} == \frac{\pi}{2}\frac{\nu}{U_e \delta} \quad [3.546]$$

The von Kármán equation is then written as:

$$0.137 \frac{d\delta}{dx} = \frac{\pi}{2}\frac{\nu}{U_e \delta} \quad [3.547]$$

$$\delta\frac{d\delta}{dx}=11.47\frac{\nu}{U_e\delta}\qquad[3.548]$$

$$\frac{d\delta^2}{dx}=22.93\frac{\nu}{U_e\delta}\quad;\quad x=0\ ;\ \delta=0\qquad[3.549]$$

From this simple differential equation associated with its boundary condition, it appears that:

$$\delta^2=22.93\frac{\nu x}{U_e}\qquad[3.550]$$

$$\delta=4.79\sqrt{\frac{\nu x}{U_e}}\qquad[3.551]$$

or again, by introducing the axial Reynolds number $R_x=\frac{U_e x}{\nu}$, [3.552]

$$\frac{\delta}{x}=4,79\sqrt{\frac{\nu x}{U_e}}=\frac{4.79}{\sqrt{R_x}}\qquad[3.553]$$

The friction coefficient C_f is then:

$$\frac{C_f}{2}=\frac{\pi}{2}\frac{\nu}{U_e\delta}=0.328\sqrt{\frac{\nu}{U_e x}}=\frac{0.328}{\sqrt{R_x}}\qquad[3.554]$$

These results are to be compared with those of Blasius. The expressions of $\frac{\delta}{x}$ and C_f have forms that are completely identical to those of Blasius.

By adopting this third profile, we underestimate the boundary layer thickness by 3.2% and the friction coefficient by 1.2%. Another very good approximation is validated here, which uses circular functions in the calculation.

2) Thermal problem

The velocity profiles are assigned the following form:

For $y \leq \delta$; $\dfrac{u}{U_e} = \dfrac{3y}{2\delta} - \dfrac{1}{2}\left(\dfrac{y}{\delta}\right)^3$ [3.555]

For $y > \delta$; $\dfrac{u}{U_e} = 1$ [3.556]

The reduced temperature profiles are assigned the following form:

For $y \leq \delta$; $\theta = \dfrac{3y}{2\delta_{th}} - \dfrac{1}{2}\left(\dfrac{y}{\delta_{th}}\right)^3$ [3.557]

For $y > \delta$; $\theta = 1$ [3.558]

We then define a "thermal" thickness:

$$\delta_\theta = \int_0^\infty (1-\theta)\frac{u}{U_e}dy \quad [3.559]$$

To simplify things, let us assume that the two boundary layer thicknesses are equal: $\delta_{th} = \delta$.

The thermal thickness is then written as:

$$\delta_\theta = \int_0^\delta \left(1 - \frac{3y}{2\delta_{th}} + \frac{1}{2}\left(\frac{y}{\delta_{th}}\right)^3\right)\left[\frac{3y}{2\delta} - \frac{1}{2}\left(\frac{y}{\delta}\right)^3\right]$$

$$dy = \int_0^\delta \left(1 - \frac{3y}{2\delta} + \frac{1}{2}\left(\frac{y}{\delta}\right)^3\right)\left[\frac{3y}{2\delta} - \frac{1}{2}\left(\frac{y}{\delta}\right)^3\right]\frac{\delta}{\delta}dy$$

We can then calculate δ_θ as a function of δ; let us apply the variable change:

$$\xi = \frac{y}{\delta} \quad [3.560]$$

$$\delta_\theta = \delta\int_0^\delta \left(1 - \frac{3\xi}{2} + \frac{1}{2}\xi^3\right)\left[\frac{3\xi}{2} - \frac{1}{2}\xi^3\right]d\xi \quad [3.561]$$

$$\delta_\theta = \delta \int_0^\delta \left(\frac{3\xi}{2} - \frac{1}{2}\xi^3 - \frac{9\xi^2}{4} + \frac{3}{4}\xi^4 + \frac{3\xi^4}{4} - \frac{1}{4}\xi^6 \right) d\xi \qquad [3.562]$$

$$\delta_\theta = \delta \left[\frac{3\xi^2}{4} - \frac{\xi^4}{8} - \frac{9\xi^3}{12} + \frac{3\xi^5}{20} + \frac{3\xi^5}{20} - \frac{\xi^7}{28} \right]_0^1 \qquad [3.563]$$

Ultimately:

$$\delta_\theta = 0.14\,\delta \qquad [3.564]$$

The Stanton is then expressed as:

$$St_x = \frac{d\,\delta_\theta}{d\,x} = 0.14 \frac{d\,\delta}{d\,x} \qquad [3.565]$$

with:

$$\delta = 4.92 \sqrt{\frac{\nu x}{U_e}} \qquad [3.566]$$

We then get:

$$St_x = 0.14 \frac{d\,\delta}{d\,x} = 0.14 * 4.92 \frac{1}{2} \sqrt{\frac{\nu}{U_e\,x}} \qquad [3.567]$$

$$St_x = 0.344\, R_x^{-0.5} \qquad [3.568]$$

This value is close to that obtained by the analytical method for a Prandtl equal to the unit:

$$St_x = 0.332\, R_x^{-0.5} \qquad [3.569]$$

EXAMPLE 3.9.– Parietal temperature step: Eckert's problem

We consider a **boundary layer developed by** a uniform velocity flow U_e, without a longitudinal pressure gradient **on a plate that is parallel to** the current lines of this potential flow.

This flow is at a far temperature that is uniform, T_e. We will use P_r as the Prandtl number of the fluid, the other classical notations remaining the same: $\rho, \mu, \lambda, C_P, a$.

The plate is heated to a constant temperature T_W from just one abscissa x_0. The existence of a length of unheated plate on x_0, from the leading edge, characterizes the temperature step phenomenon.

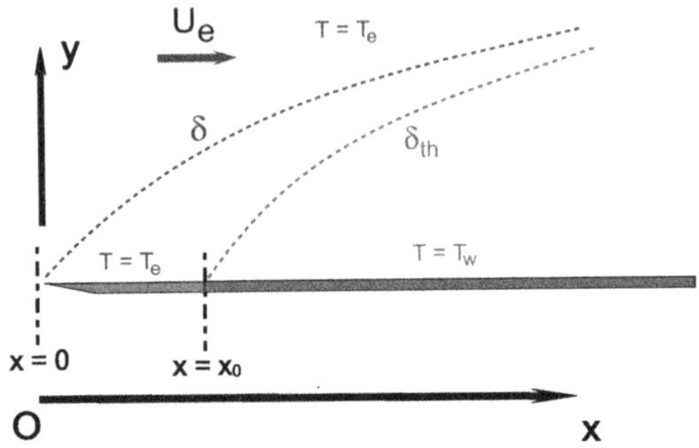

Figure 3.10. *Boundary layer with parietal temperature step. For a color version of this figure, see www.iste.co.uk/ledoux/heat3.zip*

A thermal boundary layer of thickness $\delta_{th}(x)$ then develops from x_0.

In this flow geometry, Eckert proposed the following expression for the Stanton:

$$St(x,x_0) = g\left(\frac{x}{x_0}\right) St(x, x_0 = 0) \quad [3.570]$$

$$\text{So: } St(x,x_0) = g\left(\frac{x}{x_0}\right) 0.332\, R_x^{-0.5}\ \text{Pr}^{-2/3} \quad [3.571]$$

$$\text{with: } g\left(\frac{x_0}{x}\right) = \left[1-\left(\frac{x_0}{x}\right)^{3/4}\right]^{-1/3} \quad [3.572]$$

We suggest trying to find this result using an integral method, by calculating the parietal flow density from x_0, $\varphi_W(x,x_0)$, $x > x_0$.

The physical properties of the fluid being constant, the dynamic and thermal problems are uncoupled. We will therefore assume that the Blasius results are applicable to the dynamic boundary layer. Furthermore, it will be assumed that a good approximation of the reduced velocity profile in the region of interest of the thermal boundary layer will be given by:

$$\frac{u}{U_e} = \frac{y}{\delta} \quad [3.573]$$

where $\delta(x)$ is the thickness of the dynamic boundary layer.

We will use an integral method here. We define a reduced temperature θ by:

$$\theta = \frac{T(x,y) - T_e}{T_W - T_e} \quad [3.574]$$

The reduced velocity and temperature profiles are approximated by:

$$\begin{aligned} \frac{u}{U_e} &= \frac{y}{\delta} \; ; y < \delta \\ \frac{u}{U_e} &= 1 \; ; \; y \geq \delta \end{aligned} \quad [3.575]$$

$$\begin{aligned} \theta &= \frac{y}{\delta_{th}} \; ; \; y < \delta_{th} \\ \theta &= \frac{y}{1_{th}} \; ; \; y \geq \delta_{th} \end{aligned} \quad [3.576]$$

We define $\Delta(x) = \dfrac{\delta_{th}(x)}{\delta(x)}$.

We will assume, in this problem, that we are in a case where $\Delta(x) < 1$.

We will use a result found in Example 3.8:

$$\frac{\delta(x)}{x} = \frac{\sqrt{12}}{R_x^{0.5}} \quad [3.577]$$

1) Calculate δ_θ as a function of δ and Δ. Let us recall that, inevitably, $\Delta \leq 1$.

2) Calculate $St(x)$ as a function of δ_{th}, then of δ and Δ.

Using the integrated equation in **1**, as well as the known form of $\delta(x)$, find the differential equation verified by $\Delta^3(x)$.

3) We had to find an equation of the following form in **2**:

$$x\frac{d\Delta^3}{dx} + \alpha\Delta^3 = \beta \qquad [3.578]$$

where α and β are two constants.

Solve the equation and give the expression of $\Delta(x, x_0)$.

NOTE.– It is useful to remember that the solution of such an equation is the sum of the general solution of the equation without a second member, and a particular solution of the complete equation.

4) Deduce the expression of the Stanton number as a function of x and x_0.

What happens to this expression when there is an absence of scale ($x_0 = 0$)?

Compare the expression obtained to a classical result.

SOLUTION TO EXAMPLE 3.9.–

1) We define $\Delta(x)$**:** $\delta_{th}(x) = \Delta(x)\,\delta(x)$.

Example **3.8** showed us that, using a linear velocity profile in the boundary layer, the integral method gives us the following for the dynamic boundary layer thickness $\delta(x)$:

$$\frac{\delta(x)}{x} = \sqrt{12}\,R_x^{-0.5} \qquad [3.579]$$

Adopting a linear profile for the reduced temperature profile:

$$\theta(y,x)=\frac{y}{\delta(x)} \quad [3.580]$$

For $y<\delta_{th}(x)$.

And beyond this:

$$\theta(y,x)=1 \quad [3.581]$$

The thermal thickness will be written as:

$$\delta_\theta(x)=\int_0^\infty (1-\theta(y,x))\frac{u}{U_e}dy=\int_0^{\delta_{th}}\left(1-\frac{y}{\delta_{th}}\right)\frac{y}{\delta}dy \quad [3.582]$$

The integration is done between $y=0$ and $y=\delta_{th}$, because the integrand is zero for $y>\delta_{th}$.

Let us calculate $\delta_\theta(x)$ by introducing $\Delta(x)$, and performing the change of variable $\xi=\frac{y}{\delta_{th}}$ in the integral:

$$\begin{aligned}\delta_\theta(x)&=\int_0^{\delta_{th}}\left(1-\frac{y}{\delta_{th}}\right)\frac{y}{\delta}dy\\&=\int_0^{\delta_{th}}\left(1-\frac{y}{\delta_{th}}\right)\frac{\Delta\, y}{\delta_{th}}\delta_{th}d\frac{y}{\delta_{th}}=\delta_{th}(x)\Delta(x)\int_0^1(1-\xi)\xi\, d\xi\end{aligned} \quad [3.583]$$

We get: $\int_0^1(1-\xi)\xi\, d\xi=\int_0^1\left(\xi-\xi^2\right)d\xi=\left.\frac{\xi^2}{2}-\frac{\xi^3}{3}\right|_0^1=\frac{1}{6}$ [3.584]

So, finally:

$$\delta_\theta(x)=\frac{\Delta(x)\delta_{th}(x)}{6}=\frac{\Delta^2(x)\delta(x)}{6} \quad [3.585]$$

2) The Stanton number can be written in two ways.

From its definition, and taking into account the definition of $\theta = \dfrac{(T - T_w)}{(T_e - T_w)}$ and that of thermal diffusivity $a = \dfrac{\lambda}{\rho c_p}$:

$$St\,x = \frac{\varphi_w}{\rho c_p U_e (T_e - T_w)} = \frac{\lambda \left.\dfrac{\partial T}{\partial y}\right|_{y=0}}{\rho c_p U_e (T_e - T_w)} = \frac{\lambda \left.\dfrac{\partial (T_e - T_w)\theta}{\partial y}\right|_{y=0}}{\rho c_p U_e (T_e - T_w)} = \frac{a \left.\dfrac{\partial \theta}{\partial y}\right|_{y=0}}{U_e}$$

Given the linear temperature profile:

$$\left.\frac{\partial \theta}{\partial y}\right|_{y=0} = \left.\frac{\partial}{\partial y}\frac{y}{\delta_{th}}\right|_{y=0} = \frac{1}{\delta_{th}} \qquad [3.586]$$

$$St\,x = \frac{a \left.\dfrac{\partial \theta}{\partial y}\right|_{y=0}}{U_e} = \frac{a}{U_e\,\delta_{th}} = \frac{a}{U_e\,\Delta\,\delta} \qquad [3.587]$$

The integrated energy equation gave us:

$$St = \frac{d\,\delta_\theta}{d\,x} \qquad [3.588]$$

Equating the two expressions of the Stanton number, we obtain the differential equation desired:

$$St = \frac{d\,\delta_\theta}{d\,x} = \frac{d}{d\,x}\frac{\Delta^2\,\delta}{6} = \frac{a}{U_e\,\Delta\,\delta} \qquad [3.589]$$

which we can put in the following form:

$$\Delta\,\delta \frac{d}{d\,x}\frac{\Delta^2\,\delta}{6} = \frac{a}{U_e} \qquad [3.590]$$

The boundary condition will be given in $x = x_0$, where the thermal boundary layer begins; its thickness $\delta_{th} = \Delta(x)\,\delta(x)$ is therefore still zero:

$$x = x_0 \quad ; \quad \Delta(x_0) = 0 \qquad [3.591]$$

$$\Delta\delta \frac{d}{dx}\frac{\Delta^2\,\delta}{6} = \frac{a}{U_e} \qquad [3.592]$$

$$\delta(x) = \sqrt{\frac{12\,\nu x}{U_e}} = 3.46\sqrt{\frac{\nu x}{U_e}} \qquad [3.593]$$

$$St\,x = \frac{a\left.\dfrac{\partial\theta}{\partial y}\right|_{y=0}}{U_e} = \frac{a}{U_e\,\delta_{th}} = \frac{a}{U_e\,\Delta\,\delta} \qquad [3.594]$$

3) To find $\Delta(x, x_0)$, we must therefore solve the differential equation:

$$\Delta\delta \frac{d}{dx}\frac{\Delta^2\,\delta}{6} = \frac{a}{U_e} \qquad [3.595]$$

So:

$$\frac{1}{6}\left[\Delta^3\,\delta\,\frac{d}{dx}\delta + 2\,\delta^2\,\Delta^2\,\frac{d}{dx}\Delta\right] = \frac{a}{U_e} \qquad [3.596]$$

$\delta(x)$ is of the form $\delta(x) = A\sqrt{x}$

with $A = 4.92\sqrt{\dfrac{\nu}{U_e}}$ The equation is rewritten as:

$$\Delta^3\,\frac{1}{2} + 2\,x\,\frac{d}{dx}\frac{\Delta^3}{3} = \frac{6a}{A^2 U_e} \qquad [3.597]$$

So:

$$x \frac{d}{dx}\Delta^3 + \frac{3}{4}\Delta^3 = \frac{9a}{A^2 U_e} \qquad [3.598]$$

which is of the form given in Question 5:

$$x \frac{d\,\Delta^3}{dx} + \alpha\,\Delta^3 = \beta \qquad [3.599]$$

with:

$$\alpha = \frac{3}{4} \qquad [3.600]$$

$$\beta = \frac{9a}{A^2\,U_e} = \frac{9a}{4{,}92\frac{\nu}{U_e}\,U_e} = 1.83\,\mathrm{Pr}^{-1} \qquad [3.601]$$

Let us solve the equation in Δ^3:

$$x \frac{d\,\Delta^3}{dx} + \alpha\,\Delta^3 = -\beta \qquad [3.602]$$

The general solution in Δ^3 will be the sum of a general solution of the equation without a second member, and a particular solution of the complete equation.

For the equation without a second member:

$$x \frac{d\,\Delta^3}{dx} + \alpha\,\Delta^3 = 0 \qquad [3.603]$$

$$\frac{d\,\Delta^3}{\Delta^3} = \frac{-\alpha\,dx}{x} \qquad [3.604]$$

$$Ln\,\Delta^3 = -\alpha\,Ln\,x + Ln\,C \qquad [3.605]$$

$$\Delta^3 = -C\,x^{\alpha} \qquad [3.606]$$

For the complete equation:

$$\Delta^3 = -\frac{\beta}{\alpha} \qquad [3.607]$$

is a solution with zero derivative, and is satisfactory.

Δ^3 will thus have the general form:

$$\Delta^3 = -C\,x^{\alpha} - \frac{\beta}{\alpha} \qquad [3.608]$$

which is determined by the boundary condition:

$$x = x_0 \quad ; \quad \Delta^3 = -Cx_0^{\alpha} - \frac{\beta}{\alpha} \qquad [3.609]$$

$$C = -\frac{\beta}{\alpha x_0^{\alpha}} \qquad [3.610]$$

and finally:

$$\Delta^3 = \frac{\beta}{\alpha x_0^{\alpha}}\,x^{\alpha} - \frac{\beta}{\alpha} = \frac{\beta}{\alpha}\left(\frac{x}{x_0}\right)^{\alpha} \qquad [3.611]$$

$$\frac{\beta}{\alpha} = \frac{3}{4} * \frac{4}{3}\,\mathrm{Pr}^{-1} = \mathrm{Pr}^{-1} \qquad [3.612]$$

So, ultimately:

$$\Delta^3 = \frac{\beta}{\alpha x_0^{\alpha}}\,x^{\alpha} - \frac{\beta}{\alpha} = \frac{\beta}{\alpha}\left(\frac{x}{x_0}\right)^{\alpha} \qquad [3.613]$$

That is:

$$\frac{\delta_{th}}{\delta} = \Delta = \mathrm{Pr}^{-1/3}\left[\left(\frac{x}{x_0}\right)^{3/4} - 1\right]^{1/3} \qquad [3.614]$$

When $x_0 = 0$, a known result is found:

$$\frac{\delta_{th}}{\delta} \Pr^{-1/3} \tag{3.615}$$

4) With the chosen reduced temperature profile form, we have:

$$\varphi_W = \lambda \left.\frac{\partial T}{\partial y}\right|_{y=0} = \lambda \left(T_e - T_W\right) \left.\frac{\partial \theta}{\partial y}\right|_{y=0} = \lambda \left(T_e - T_W\right) \frac{1}{\delta_{th}} \tag{3.616}$$

$$\varphi_W = \lambda \left(T_e - T_W\right) \frac{1}{\Delta \delta} \tag{3.617}$$

$$\varphi_W = \lambda \left(T_e - T_W\right) \Pr^{1/3} \left[\left(\frac{x}{x_0}\right)^{3/4} - 1\right]^{-1/3} \sqrt{\frac{U_e}{12 \nu x}} \tag{3.618}$$

$$\varphi_W = \lambda \left(T_e - T_W\right) \Pr^{1/3} \left[\left(\frac{x}{x_0}\right)^{3/4} - 1\right]^{-1/3} 0.289 \sqrt{\frac{U_e}{\nu x}} \tag{3.619}$$

The Stanton comes to:

$$St\left(x, x_0\right) = \frac{\varphi_W}{\rho c_P U_e \left(T_e - T_W\right)} = \frac{\lambda}{\rho c_P} \Pr^{1/3} \left[\left(\frac{x}{x_0}\right)^{3/4} - 1\right]^{-1/3} \frac{1}{U_e} \sqrt{\frac{U_e}{12 \nu x}} \tag{3.620}$$

Noting that: $\dfrac{\lambda}{\rho c_P} = a = \dfrac{\nu}{\Pr}$

It comes to:

$$St\left(x, x_0\right) = 0.289 \left[\left(\frac{x}{x_0}\right)^{3/4} - 1\right]^{-1/3} R_x^{-0.5} \Pr^{-1/3} \tag{3.621}$$

which is, in fact, of the following form:

$$St(x, x_0) = g\left(\frac{x}{x_0}\right) St(x, x_0 = 0) \qquad [3.622]$$

with: $$g\left(\frac{x_0}{x}\right) = \left[1 - \left(\frac{x_0}{x}\right)^{3/4}\right]^{-1/3} \qquad [3.623]$$

EXAMPLE 3.10. Plate in the path of the wind. Friction and heat transfer with parietal fluid extraction. Integral method approach

We resume the problem set in section 3.1, which was treated by analytical approach

We suggest studying the boundary layer produced by a uniform flow with distanced velocity U_e, and a porous flat plate placed in the path of the wind, parallel to the current lines of the potential flow. The fluid is incompressible, with density ρ and dynamic viscosity μ. All these physical properties are constant.

The fluid is drawn along the whole length of the plate, with a constant velocity of module W. The flow, being a plane flow, will be marked by a system of Oxy axes (Ox following the plate). The origin O is chosen at the leading edge. The velocity will have two components, u and v respectively along Ox and Oy. It will be assumed that from a certain abscissa x_0, the flow is self-similar, that is, the longitudinal component of the local velocity solely depends on the distance to the wall, y: $u = u(y)$. The zone near the leading edge ($x < x_0$) will be called the non-similar zone.

1) Address Question **1.1 of the original problem**, if it has not already been done.

2) For this problem, establish the relation between δ_2, C_f and W. We will establish this relation for the whole boundary layer, including the non-similar zone. Let us recall that, in this zone, $u = u(x, y)$; $v = v(x, y)$.

The integral method will be used to determine the friction coefficient C_f in the self-similar zone.

To do this, it will be necessary to once more take the previously-recalled analysis, in order to establish the relation of von Kármán, taking into account the fact that v is no longer zero at the wall. It will be assumed that:

The thickness of the dynamic boundary layer δ is constant in this zone.

The profile of $u(y)$ is self-similar:

$$\frac{u}{U_e} = f\left(\frac{y}{\delta}\right) \quad [3.624]$$

$f\left(\frac{y}{\delta}\right)$ being a function that we will not have to give ourselves a priori.

Compare the expression of $C_f(x)$ found to the one determined in Question 1.

3) Establish the integral relation between $S_t(x)$, δ^* and W, where δ^* is a thermal thickness defined by:

$$\delta^* = \int_0^\infty \frac{u}{U_e}(1-\theta)\,dy \quad [3.625]$$

where θ is the reduced variable:

$$\theta = \frac{T - T_W}{T_e - T_W} \quad [3.626]$$

It will be assumed that the profile of θ is self-similar, and that its form only depends on P_r:

$$\theta = g\left(\frac{y}{\delta_{th}}, P_r\right) \quad [3.627]$$

g is an initially unknown function, whose form we will not have to determine in this problem.

The integral relation for the entire boundary layer, including the non-similar zone ($0 < x < x_0$), will be established.

4) Find the Stanton expression in the similar zone. Compare it to the result of the analytical approach.

SOLUTION TO EXAMPLE 3.10.–

Integral method

The integral method has been presented in **3.2** and in the previous examples. In the present case, the analysis will be a little more complicated. The von Kármán expression is not directly applicable. We have to repeat this analysis here.

1) Let us recall that in section **3.1**, we established that the lateral component of the velocity is constant in the boundary layer:

$$v = -W \quad [3.628]$$

2) Integral method for an external boundary layer, in Cartesian coordinates:

$$\frac{\partial u}{\partial x} + \frac{\partial v}{\partial y} = 0 \quad [3.629]$$

$$u\frac{\partial u}{\partial x} + v\frac{\partial u}{\partial y} = \nu\frac{\partial^2 u}{\partial y^2} \quad [3.630]$$

Let us integrate the continuity and momentum equations on y between 0 and infinity:

The first one gives us:

$$\int_0^\infty \frac{\partial u}{\partial x} dy = -\int_0^\infty \frac{\partial v}{\partial y} dy = -\left[v_\infty - v(y=0)\right] = W - v_\infty \quad [3.631]$$

v_∞ is not zero (external flow deviation), due to the mass flow deficit in the boundary layer.

$$\int_0^\infty \left(u\frac{\partial u}{\partial x} + v\frac{\partial u}{\partial y}\right) dy = \nu\frac{\partial u}{\partial y}\Big|_0^\infty = -\frac{1}{\rho}\mu\frac{\partial u}{\partial y} = \frac{-\tau_w}{\rho} \quad [3.632]$$

where τ_w is the parietal stress.

We define the friction coefficient C_f:

$$C_f = \frac{2\tau_w}{\rho U_e^2} \quad [3.633]$$

$$\int_0^\infty \left(u\frac{\partial u}{\partial x} + v\frac{\partial u}{\partial y} \right) dy = v\frac{\partial u}{\partial y}\Big|_0^\infty = -\frac{1}{\rho}\mu\frac{\partial u}{\partial y} = \frac{-\tau_w}{\rho}$$

$$\int_0^\infty \left(u\frac{\partial u}{\partial x} + v\frac{\partial u}{\partial y} \right) dy = \int_0^\infty \left(\frac{\partial u^2}{\partial x} - u\frac{\partial u}{\partial x} + \frac{\partial uv}{\partial y} - u\frac{\partial v}{\partial y} \right) dy = \int_0^\infty \left(\frac{\partial u^2}{\partial x} + \frac{\partial uv}{\partial y} \right) dy$$

As $-u\frac{\partial u}{\partial x} - u\frac{\partial v}{\partial y} = -u\left(\frac{\partial u}{\partial x} + \frac{\partial v}{\partial y} \right)$ is zero because of the continuity equation.

Furthermore:

$$\int_0^\infty \frac{\partial uv}{\partial y} dy = U_e v_\infty = U_e \left(W - \int_0^\infty \frac{\partial u}{\partial x} dy \right) \quad [3.634]$$

$$\begin{aligned} &\int_0^\infty \left(u\frac{\partial u}{\partial x} + v\frac{\partial u}{\partial y} \right) dy = \\ &= \int_0^\infty \left(\frac{\partial u^2}{\partial x} + \frac{\partial uv}{\partial y} \right) dy = \int_0^\infty \left(\frac{\partial u^2}{\partial x} - U_e\frac{\partial u}{\partial x} \right) dy + W U_e = \frac{\tau_w}{\rho} \end{aligned} \quad [3.635]$$

$$\int_0^\infty \left(\frac{\partial u^2}{\partial x} - U_e\frac{\partial u}{\partial x} \right) dy = \frac{d}{dx}\int_0^\infty \left(u^2 - U_e u \right) dy = \frac{\tau_w}{\rho} - W U_e \quad [3.636]$$

These expressions are scaled by dividing the two terms of the equation by U_e^2. We then define a momentum thickness:

$$\delta_2 = \int_0^\infty \left(1 - \frac{u}{U_e} \right) \frac{u}{U_e} dy \quad [3.637]$$

and we introduce the friction coefficient:

$$C_f = \frac{2\tau_W}{\rho U_e^2} \quad [3.638]$$

Finally, the form of the von Kármán equation that is specific to our problem shows itself:

$$\frac{d\delta_2}{dx} = \frac{C_f}{2} - \frac{W}{U_e} \quad [3.639]$$

Let us establish the relationship between δ and δ_2, using its previous definition:

$$\delta_2 = \int_0^\infty \left(1 - \frac{u}{U_e}\right)\frac{u}{U_e}dy = \delta \int_0^\infty \left(1 - f\left(\frac{y}{\delta}\right)\right) f\left(\frac{y}{\delta}\right) d\frac{y}{\delta} = F\,\delta \quad [3.640]$$

$$F = \int_0^\infty \left(1 - f\left(\frac{y}{\delta}\right)\right) f\left(\frac{y}{\delta}\right) d\frac{y}{\delta} \quad [3.641]$$

F is a definite integral, the value of which, as we will see, is not necessary to know.

Taking into account the invariability of δ ***in the self-similar zone***, the von Kármán law is written as:

$$\frac{d\delta_2}{dx} = F\frac{d\delta}{dx} = 0 = \frac{C_f}{2} - \frac{W}{U_e} \quad [3.642]$$

As a result, it is clear that:

$$C_f = \frac{2W}{U_e} \quad [3.643]$$

We find here that the result was already established by the analytical approach.

3) Let us define the thermal thickness:

$$\delta_\theta = \int_0^\infty \frac{u}{U_e}(1-\theta)\,d\,y \qquad [3.644]$$

It is assumed that the profile of θ is self-similar, and that its form only depends on P_r:

$$\theta = g\left(\frac{y}{\delta_{th}}, P_r\right) \qquad [3.645]$$

The thermal thickness becomes:

$$\delta_\theta = \int_0^\infty \frac{u}{U_e}\left(1 - g\left(\frac{y}{\delta_{th}}, P_r\right)\right)d\,y \qquad [3.646]$$

which gives us:

$$\delta_\theta = G\,\delta \qquad [3.647]$$

where $\delta_\theta = G\,\delta$ is a constant that depends on the form of the chosen profiles. We will see that we will not have to know it in the similarity zone.

Let us remember that we must distinguish δ_θ between the thickness of the thermal boundary layer δ_{th}, where $\theta = 0.99$.

For the energy equation, we will get the following with a reduced temperature for an unknown function:

$$u\frac{\partial\theta}{\partial x} + v\frac{\partial\theta}{\partial y} = a\frac{\partial^2\theta}{\partial y^2} \qquad [3.648]$$

$$0 - W\frac{\partial\theta}{\partial y} = v\frac{\partial^2\theta}{\partial y^2} \qquad [3.649]$$

which we integrate into y:

$$\int_0^\infty W\frac{\partial\theta}{\partial y}\,d\,y = \int_0^\infty a\frac{\partial^2\theta}{\partial y^2}\,d\,y \qquad [3.650]$$

$$\int_0^\infty \left(u\frac{\partial \theta}{\partial x} + v\frac{\partial \theta}{\partial y} \right) dy = = \int_0^\infty \left[\left(\frac{\partial u\theta}{\partial x} + \frac{\partial v\theta}{\partial y} \right) - \theta \left(\frac{\partial u}{\partial x} + \frac{\partial v}{\partial y} \right) \right] dy \qquad [3.651]$$

Due to the continuity equation:

$$\int_0^\infty \left(u\frac{\partial \theta}{\partial x} + v\frac{\partial \theta}{\partial y} \right) dy = = \int_0^\infty \left(\frac{\partial u\theta}{\partial x} + \frac{\partial v\theta}{\partial y} \right) dy \qquad [3.652]$$

Let us add the integrated second member of the energy equation:

$$\int_0^\infty \left(\frac{\partial u\theta}{\partial x} + \frac{\partial v\theta}{\partial y} \right) dy = \int_0^\infty a \frac{\partial^2 \theta}{\partial y^2} dy \qquad [3.653]$$

$$\frac{\partial}{\partial x} \int_0^\infty u\theta \, dy + v\theta \Big|_0^\infty = a \frac{\partial \theta}{\partial y}\Big|_0^\infty \qquad [3.654]$$

We remember that:

$$\int_0^\infty \frac{\partial v}{\partial y} dy = -\int_0^\infty \frac{\partial u}{\partial x} d\, y \qquad [3.655]$$

So:

$$v_\infty - W = -\int_0^\infty \frac{\partial u}{\partial x} d\, y \qquad [3.656]$$

Note that the suction speed W is negative.

Taking the integrated equation again, and noting that:

$$v\theta \Big|_0^\infty = v_\infty * 1 - W * 0 = v_\infty \qquad [3.657]$$

$$a \frac{\partial \theta}{\partial y}\Big|_0^\infty = \frac{\partial}{\partial x} \int_0^\infty u\theta \, dy + v_\infty = \frac{\partial}{\partial x} \int_0^\infty u\theta \, dy - W - \int_0^\infty \frac{\partial u}{\partial x} d\, y \qquad [3.658]$$

$$\frac{\partial}{\partial x}\int_0^\infty u\theta\,dy+W-\int_0^\infty \frac{\partial u}{\partial x}d\,y=\frac{\partial}{\partial x}\int_0^\infty (u\theta-u\,)dy-W \qquad [3.659]$$

$$a\left.\frac{\partial\,\theta}{\partial\,y}\right|_0^\infty=\frac{1}{\rho c_P(T_e-T_W)}\left(-\lambda\left.\frac{\partial T}{\partial\,y}\right|_{y=0}\right)=\frac{-\varphi_W}{\rho c_P(T_e-T_W)} \qquad [3.660]$$

Ultimately:

$$\frac{\varphi_W}{\rho c_P(T_e-T_W)}=\frac{\partial}{\partial x}\int_0^\infty u(1-\theta\,)dy-W \qquad [3.661]$$

Dividing the two terms by U_e:

$$\frac{\varphi_W}{\rho c_P(T_e-T_W)}=\frac{\partial}{\partial x}\int_0^\infty \frac{u}{U_e}(1-\theta\,)dy-\frac{W}{U_e} \qquad [3.662]$$

$$\frac{\varphi_W}{\rho c_P(T_e-T_W)}=\frac{\partial}{\partial x}\delta_{th}-\frac{W}{U_e} \qquad [3.663]$$

In the zone where δ is constant:

$$\frac{d\,\delta_\theta}{d\,x}=G\frac{d\,\delta}{d\,x}=0 \qquad [3.664]$$

Taking the absolute value of W, it remains that:

$$\frac{\varphi_W}{\rho c_P(T_e-T_W)}=\frac{W}{U_e} \qquad [3.665]$$

This is the result found by the analytical method.

EXAMPLE 3.11.– Turbulent boundary layer. Use of the "$\frac{1}{7}$" law.

We want to apply the integral method to an established turbulent boundary layer, assuming the following forms for the velocity and reduced temperature profiles:

For $y \le \delta$; $\dfrac{u}{U_e} = \left(\dfrac{y}{\delta}\right)^{1/7}$ [3.666]

For $y > \delta$; $\dfrac{u}{U_e} = 1$ [3.667]

For $y \le \delta$; $\theta = \left(\dfrac{y}{\delta}\right)^{1/7}$ [3.668]

For $y > \delta$; $\theta = 1$ [3.669]

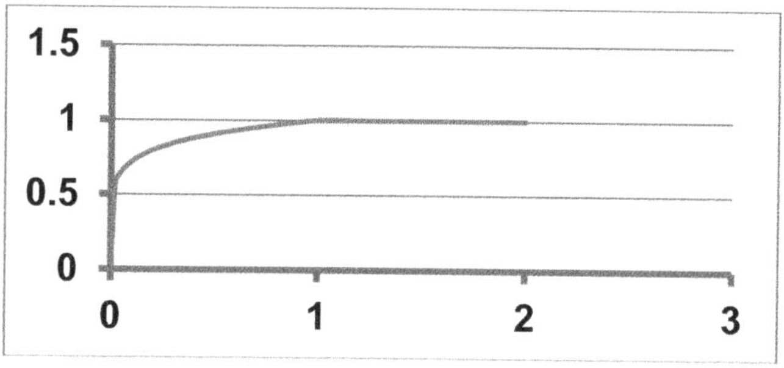

Figure 3.11. *The "$\frac{1}{7}$" profile*

1) Determine the expression for the ratio of momentum thickness to boundary layer thickness. Determine the expression of the friction coefficient. Conclusion.

2) We place ourselves in the case where the dynamic and thermal boundary layers are of equal thickness.

Determine the expression for the ratio of thermal thickness to boundary layer thickness. Determine the expression of the Stanton. Conclusion.

SOLUTION TO EXAMPLE 3.11.–

1) We know the expression of von Kármán:

$$\frac{d\,\delta_2}{d\,x} = \frac{C_f}{2} \quad [3.670]$$

where:

$$\delta_2 = \int_0^\infty \left(1 - \frac{u}{U_e}\right)\frac{u}{U_e}dy \quad [3.671]$$

$$C_f = \frac{2\tau_w}{\rho U_e^2} \quad [3.672]$$

The method is applied in the same way for the laminar case:

The momentum thickness is calculated:

$$\delta_2 = \int_0^\infty \left(1 - \frac{u}{U_e}\right)\frac{u}{U_e}dy = \int_0^1 \left(1 - \left(\frac{y}{\delta}\right)^{1/7}\right)\left(\frac{y}{\delta}\right)^{1/7} \delta\, d\frac{y}{\delta} = \int_0^1 \left(1 - \xi^{1/7}\right)\xi^{1/7}\, \delta\, d\xi \quad [3.673]$$

The integral has an upper limit, given that the integrand is zero beyond it.

$$\int_0^1 \left(1 - \xi^{1/7}\right)\xi^{1/7}\, \delta\, d\xi = \int_0^1 \left(\xi^{1/7} - \xi^{2/7}\right)d\xi = \frac{7}{8}\xi^{8/7} - \frac{7}{9}\xi^{9/7}\Big|_0^1 \quad [3.674]$$

$$\int_0^1 \left(1 - \xi^{1/7}\right)\xi^{1/7}\, \delta\, d\xi = \left(\frac{7}{8} - \frac{7}{9}\right)\delta = 0.097\,\delta \quad [3.675]$$

At the wall, we determine the parietal stress τ_w, then the C_f. Let us recall that at the wall, the friction is driven by molecular properties:

$$\tau_W = \mu \frac{\partial u}{\partial y}\Big|_{y=0} = \frac{\mu U_e}{\delta}\frac{\partial \dfrac{u}{U_e}}{\partial \dfrac{y}{\delta}}\Bigg|_{y=0} = \frac{\mu U_e}{\delta}\frac{\partial \xi^7}{\partial \xi}\Big|_{\xi=0} \quad [3.676]$$

Here, the calculation runs short:

$$\tau_W = \frac{\mu U_e}{\delta}\frac{\partial \xi^7}{\partial \xi}\Big|_{\xi=0} = \frac{\mu U_e}{\delta}7\xi^6\Big|_{\xi=0} = 0 \text{ !!!!} \quad [3.677]$$

This exercise is a trap!

Note an inevitable limit of the $\frac{1}{7}$ profile here. This form constitutes a valid approximation of the profile in the core flow, but does not bring any valid information on the slope of this profile at the wall. Let us recall that this profile was published at a time when the velocities were often measured with a Pitot tube, and could not give sufficient accuracy at the wall.

4

Heat Exchangers

4.1. Introduction and basic concepts

A heat exchanger is used when you want to transfer heat from one fluid to another, without any physical contact between them.

There are many reasons for this: health problems, chemical or safety problems, pressure problems, etc. The fluids are therefore separated by a solid wall.

It is difficult to imagine entrusting this exchange to conduction alone: conduction is a slow phenomenon, and the transfer between two reservoirs would lead to an unsteady phenomenon. The rule is therefore to proceed to an exchange between two flows. The heat exchanger will therefore initially use thermal convection.

The use of solid walls, whether tubes or plates, brings about two simultaneous transfers between a fluid and a wall, this wall carrying out the coupling. The theory that will be summarized in this chapter will therefore involve writing transfer relations and global balances. Hence the writing of local balances.

It is clear that, from here on out, the problem is particularly complex. The numerical tool allows the advanced modeling of the flows involved to be considered (at a cost which is not insignificant). The construction of heat exchangers is an old art, much older than numerical fluid mechanics. Simplified methods have been used since their inception, and are still used today, adopting a certain degree of approximation in the implementation.

These methods will be presented here.

Our objective is, once again, "cultural". The relations presented are obviously usable by the reader in the framework of small calculations (laboratory exchanger, marginal use of an exchanger). The design of industrial exchangers is at a different level of experience, and will require the use of much more complete manuals. The literature is particularly rich in this field.

There are numerous fields of application for exchangers: they vary from the domestic field (computer, automobile air-conditioning) to heavy industry fields: oil industry and electricity production.

Figure 4.1. *The world of exchangers: very different scales. For a color version of this figure, see www.iste.co.uk/ledoux/heat3.zip*

4.1.1. *Classification test*

4.1.1.1. *Exchangers technology*

Exchanger technology is particularly extensive. These devices can be classified according to different perspectives. According to the geometry of flows, the main distinction is made between the exchangers that put the following into action:

– parallel currents of the same or opposite direction;

– or other currents, such as crossed currents.

Depending on the type of fluid used:

– two fluids with similar thermal characteristics (liquid/liquid, gas/gas, etc.);

– two fluids with very different thermal characteristics (liquid/gas, etc.).

Depending on the geometry of the exchange surfaces: material wall (plate, tube, etc.).

We can also consider direct contact (air cooler, cooling tower): the problem will be solved by a study of mixing between fluids.

Here, we will only consider separate fluid exchangers, where heat transfer takes place through a wall.

4.1.1.2. *Main types of heat exchangers*

We can distinguish between

Mixer or direct-contact exchangers

Closely mixed fluid – deaerators – natural or forced convection cooling towers – steam desuperheaters.

Regenerators or batch exchangers

The exchange surface is alternately put in contact with the cold and the hot fluid: this is the case, for example, with rotary air heaters.

Continuous exchangers

The two fluids circulate continuously on either side of a separation wall, which acts as an exchange surface.

Classified by the geometry of the exchange walls

– tubular exchangers: the two fluids can be of different nature;

– plate exchangers: the exchange surfaces are flat;

– finned exchangers: the transfer is made from a fluid to a pipe. This pipe is equipped with plates that are immersed in another fluid. Note that the fins are also used with a radiative transfer.

4.2. Method of calculation of exchangers

4.2.1. *Types of exchangers*

The following models focus on two very common types of exchangers:

– co-current exchangers, where the two exchanging flows are parallel, in the same direction;

– countercurrent exchangers, where the two exchanging flows are parallel, in opposite directions.

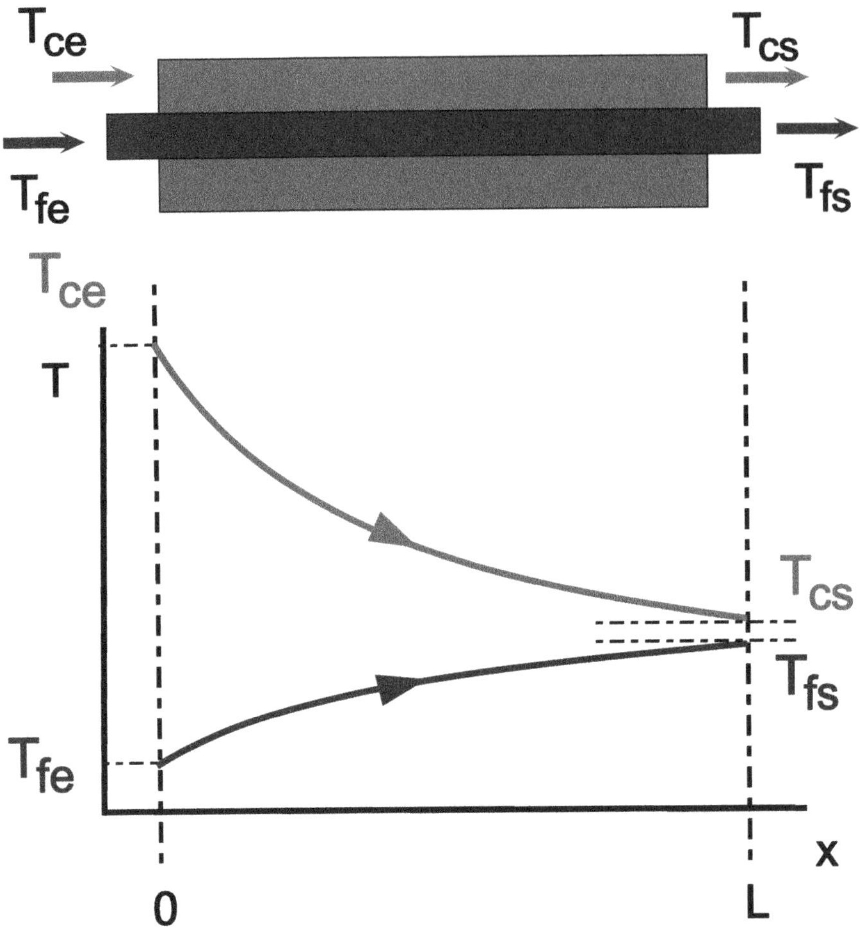

Figure 4.2. *Co-current exchanger. Diagram and temperature distributions. For a color version of this figure, see www.iste.co.uk/ledoux/heat3.zip*

These two situations are often represented by a diagram showing two coaxial tubular flows. This geometry is a very special case, and this representation does not affect the generality of the results given here.

Two methods are commonly used:

– the "Logarithmic Mean Temperature Difference" method, known as *DTLM* or ΔTLM;

– the number of transfer units method or the *NUT* method.

We will give the principle and the main elements for both co-current and countercurrent flows.

The following figures illustrate these geometries and the main thermal parameters used.

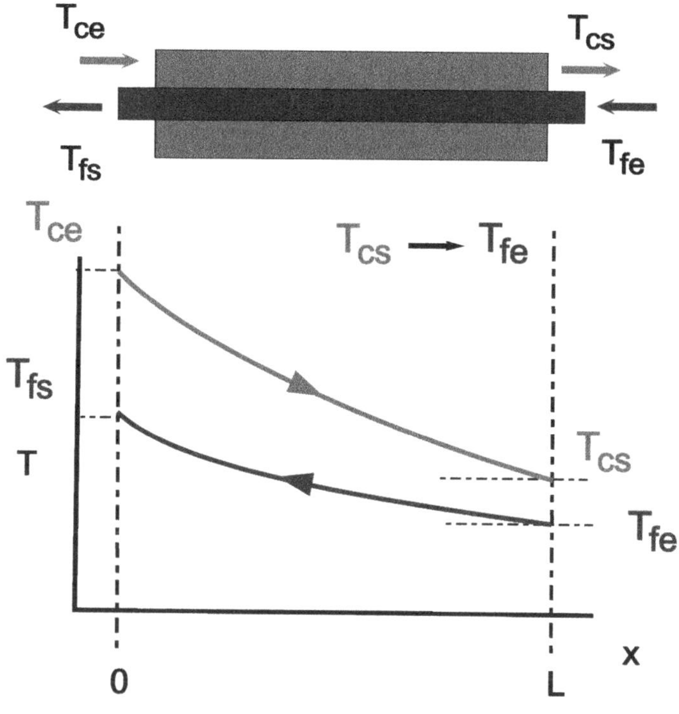

Figure 4.3. *Countercurrent exchanger. Diagram and temperature distributions. For a color version of this figure, see www.iste.co.uk/ledoux/heat3.zip*

4.2.2. Logarithmic mean temperature difference method (DTLM)

4.2.2.1. Co-current exchanger

We define:

The entry temperature of the hot fluid (“*c*” heat giver): T_{ce}.

The exit temperature of the hot fluid (“*c*” heat giver): T_{cs}.

The entry temperature of the cold fluid (“*f*” heat receiver): T_{fe}.

The exit temperature of the hot fluid (“*f*” heat receiver): T_{fs}.

The mass flow rate of the “hot” fluid: $\dot{m}_c$.

The mass flow rate of the “cold” fluid (“*f*” heat receiver): $\dot{m}_f$.

The mass heat capacity of the “hot” fluid: c_c.

The mass heat capacity of the “cold” fluid: C_f.

The “local” transfer coefficient between hot and cold fluids: k.

k is defined from the flow $d\,\Phi$ exchanged on a small length of exchanger, corresponding to an exchange surface $d\,S$:

$$d\,\Phi = k\left(T_c - T_f\right)\,d\,S$$

In the simplified theory developed here, k is a constant.

Practically speaking, k can be determined from the additivity of transfer resistances. The opposite of a global thermal resistance is then calculated by adding the following:

The convective resistance in the hot fluid, $R_{conv\,c} = \dfrac{1}{h_c}$, where h_c is the convection coefficient in the channel where the hot fluid flows.

The convective resistance in the cold fluid, $R_{conv\,f} = \dfrac{1}{h_f}$, where h_f is the convection coefficient in the channel where the cold fluid flows.

The thermal resistance of the wall separating the fluids, $R_{th\ w}$. This resistance is often complex, generally composed of a metallic wall resistance of the exchanger, as well as two fouling resistances resulting from various deposits on the walls (dust, coking, etc.) This resistance is likely to increase during the use of the exchanger.

Typical values are multiples of $10^{-4}\ m W^{-1} K$.

In total:

$$\frac{1}{k} = R_{conv\,c} + R_{conv\,f} + R_{th\,w} = \frac{1}{h_c} + \frac{1}{h_f} + R_{th\,w} \qquad [4.1]$$

NOTE.– This simple formulation is adapted to plate exchangers. It is indeed based on a one-dimensional plane conduction model. In the case of revolution surfaces, other expressions containing the radius are to be used (see parent book on conduction).

Here, we are looking for the total heat flow exchanged between the "hot" and "cold" fluids during their thermal contact in the exchanger.

A simple calorimetric consideration gives us:

$$\Phi = \dot{m}_c\, c_c\, (T_{ce} - T_{cs}) = \dot{m}_f\, c_f\, (T_{fs} - T_{fe}) = = -\dot{m}_f\, c_f\, (T_{fe} - T_{fs}) \qquad [4.2]$$

To simplify the notations, we condense the products of mass flow rates by heat capacities through two parameters, defined by:

$$q_C = \dot{m}_C\, c_C \qquad [4.3]$$

and $$q_f = \dot{m}_f\, c_f \qquad [4.4]$$

In addition, the elementary flow $d\Phi$ that is "locally" exchanged on each exchanger element of length *dl* and area *dS* can be rewritten from the same type of calorimetric balance:

$$d\Phi = k\left(T_c - T_f\right) dS = -q_c\, dT_c \ = +q_f\, dT_f \qquad [4.5]$$

From this relation, we deduce that:

$$dT_c \ = \frac{-k}{q_c}\left(T_c - T_f\right) dS \qquad [4.6]$$

$$d\,T_f = \frac{+k}{q_f}\left(T_c - T_f\right) d\,S \qquad [4.7]$$

This combines into:

$$d\left(T_c - T_f\right) = -\left[\frac{1}{q_c} + \frac{1}{q_f}\right] k\left(T_c - T_f\right) d\,S \qquad [4.8]$$

Or:

$$\frac{d\left(T_c - T_f\right)}{\left(T_c - T_f\right)} = d\,Ln\left(T_c - T_f\right) = -\left[\frac{1}{q_f} + \frac{1}{q_c}\right] k \ d\,S \qquad [4.9]$$

This is an expression that can be integrated between the entry and exit of the exchanger:

$$Ln\left(T_c - T_f\right)\Big|_e^s = -\left[\frac{1}{q_f} + \frac{1}{q_c}\right] k \ S \qquad [4.10]$$

which we will rewrite as the following, clarifying the entry and exit values:

$$Ln\frac{T_{cs} - T_{fs}}{T_{ce} - T_{fe}} = -\left[\frac{1}{q_f} + \frac{1}{q_c}\right] k \ S \qquad [4.11]$$

From the heat balance on the exchanger, we can rewrite:

$$\Phi = q_c\left(T_{ce} - T_{cs}\right) = -q_f\left(T_{fe} - T_{fs}\right) \qquad [4.12]$$

$$\left(T_{ce} - T_{cs}\right) = \frac{\Phi}{q_c} \qquad [4.13]$$

$$\left(T_{fe} - T_{fs}\right) = -\frac{\Phi}{q_f} \qquad [4.14]$$

Or even:

$$\left(T_{ce} - T_{cs}\right) - \left(T_{fe} - T_{fs}\right) = \Phi\left[\frac{1}{q_f} + \frac{1}{q_c}\right] \qquad [4.15]$$

And ultimately:

$$(T_{cs}-T_{fs})-(T_{ce}-T_{fe})=-\Phi\left[\frac{1}{q_f}+\frac{1}{q_c}\right] \quad [4.16]$$

We deduce the expression of the flow:

$$=\Phi=\frac{(T_{cs}-T_{fs})-(T_{ce}-T_{fe})}{\left[\frac{1}{q_f}+\frac{1}{q_c}\right]} \quad [4.17]$$

As:

$$Ln\frac{T_{cs}-T_{fs}}{T_{ce}-T_{fe}}=-\left[\frac{1}{q_f}+\frac{1}{q_c}\right]k\ S \quad [4.18]$$

$$\Phi=k\frac{(T_{cs}-T_{fs})-(T_{ce}-T_{fe})}{Ln\frac{T_{cs}-T_{fs}}{T_{ce}-T_{fe}}}S \quad [4.19]$$

A parameter thus appears, which has the dimension of a temperature. This parameter is designated by the logarithmic mean temperature difference, which is often noted as *DTLM*, *ΔTLM* or ΔT_{LM}, notations that we will adopt here (these are compatible with both English and French presentations). For the co-current exchanger, this *DTLM* is defined by:

$$\Delta T_{LM}=\frac{(T_{cs}-T_{fs})-(T_{ce}-T_{fe})}{Ln\frac{T_{cs}-T_{fs}}{T_{ce}-T_{fe}}} \quad [4.20]$$

And the flow is calculated by: $\Phi=k\,\Delta T_{LM}\,S$

4.2.2.2. Countercurrent exchanger

This calculation can be used for a countercurrent exchanger. We will thus find a formula with a similar structure, but ΔT_{LM} will have a different definition:

$$\Delta T_{LM} = \frac{(T_{ce} - T_{fs}) + (T_{cs} - T_{fe})}{Ln\dfrac{T_{ce} - T_{fs}}{T_{cs} - T_{fe}}} \quad [4.21]$$

We "standardize" the writing of ΔT_{LM} by introducing a temperature difference between fluids:

$$\Delta T = T_c - T_f \quad [4.22]$$

Using the entry of the hot fluid for index 1 and the exit for index 2 as reference, we define:

For a co-current exchanger:

$$\Delta T_1 = T_{ce} - T_{fe} \quad [4.23]$$

$$\Delta T_2 = T_{cs} - T_{fs} \quad [4.24]$$

For a countercurrent exchanger:

$$\Delta T_1 = T_{ce} - T_{fs} \quad [4.25]$$

$$\Delta T_2 = T_{cs} - T_{fe} \quad [4.26]$$

This results in a formal "standard" definition of ΔT_{LM}:

$$\Delta T_{LM} = \frac{\Delta T_1 - \Delta T_2}{Ln\dfrac{\Delta T_1}{\Delta T_2}} \quad [4.27]$$

And the flow is then calculated by the following unique expression:

$$\Phi = k\, \Delta T_{LM}\, S \quad [4.28]$$

In practice, we show that if ΔT_1 and ΔT_2 are sufficiently close, or more precisely, if $\Delta T_2 \geq \Delta T_1$, then ΔT_{LM} can be calculated to within 1% by the arithmetic mean of ΔT :

$$\Delta T_{LM} = \frac{\Delta T_1 + \Delta T_2}{2} \quad [4.29]$$

Practitioners use diagrams giving ΔT_{LM}, a function of ΔT_1 and ΔT_2.

4.2.3. *Number of transfer units method (N_{UT} method)*

4.2.3.1. *We are looking for the flow Φ exchanged between the two fluids*

It can be written by a heat balance on each of the fluids between entry and exit.

$$\Phi = \dot{m}_c\, c_c \left(T_{ce} - T_{cs}\right) = \dot{m}_f\, c_f \left(T_{fs} - T_{fe}\right) \quad [4.30]$$

noting that:

$$\dot{m}_f\, c_f \left(T_{fs} - T_{fe}\right) = -\dot{m}_f\, c_f \left(T_{fe} - T_{fs}\right) \quad [4.31]$$

To simplify the notations, we condense the products of mass flow rates and heat capacities through two parameters, defined by:

$$q_c = \dot{m}_c\, c_c \quad [4.32]$$

and

$$q_f = \dot{m}_f\, c_f \quad [4.33]$$

q_c and q_f are respectively relative to the “hot” (heat provider) and “cold” (heat receiver) fluid.

The writing of the total flow exchanged between “hot” and “cold” fluids is then rewritten as:

$$\Phi = q_c \left(T_{ce} - T_{cs}\right) = q_f \left(T_{fs} - T_{fe}\right) \quad [4.34]$$

NOTE.– The terms "hot" and "cold" fluid are placed in quotation marks because locally, in the reactor, the "cold" fluid can be hotter than the "hot" fluid at entry ($T_f(x) > T_{ce}$). This is particularly the case in the countercurrent reactor, where the flow is driven by the fluid.

4.2.3.2. *Driving transfer*

Depending on the relative values of q_C and q_f, there are two cases.

If $q_c < q_f$, we say that the transfer is driven by the hot fluid

This can be explained through the following reasoning:

$\Phi = q_c(T_{ce} - T_{cs}) = q_f(T_{fs} - T_{fe})$ means that

$$(T_{ce} - T_{cs}) > (T_{fs} - T_{fe}) \qquad [4.35]$$

Therefore:

$$T_{cs} - T_{fe} < T_{ce} - T_{fs} \qquad [4.36]$$

In the case of an infinitely long exchanger, we have:

$$L \to \infty \Rightarrow T_{cs} - T_{fe} \to 0\text{, so } \quad T_{cs} \to T_{fe} \qquad [4.37]$$

If $q_c < q_f$, we say that the transfer is driven by the cold fluid

This can be explained by the following reasoning:

$$\Phi = q_c(T_{ce} - T_{cs}) = q_f(T_{fs} - T_{fe}) \text{ means that } (T_{ce} - T_{cs}) < (T_{fs} - T_{fe}) \qquad [4.38]$$

Therefore:

$$(T_{ce} - T_{fs}) < (T_{cs} - T_{fe}) \qquad [4.39]$$

In the case of an infinitely long exchanger, we have:

$$L \to \infty \Rightarrow T_{ce} - T_{fs} \to 0\text{, so } \quad T_{fs} \to T_{ce} \qquad [4.40]$$

4.2.3.3. *Imbalance factor*

An imbalance factor $r = \frac{q_{\min}}{q_{\max}}$ is defined, whose use will become apparent later:

This imbalance factor can be expressed as a function of the entry and exit temperatures.

Just as before, we will distinguish the case of heat exchangers controlled by the hot or cold fluid.

If the exchange is driven by the hot fluid, then:

$$q_c < q_f \quad \Rightarrow \quad q_{\min} = q_c \quad ; \quad q_{\max} = q_f \qquad [4.41]$$

Noting that, in this case:

$$\Phi = q_c \left(T_{ce} - T_{cs}\right) = q_f \left(T_{fs} - T_{fe}\right) = q_{\min} \left(T_{ce} - T_{cs}\right) = q_{\max} \left(T_{fs} - T_{fe}\right) \qquad [4.42]$$

Therefore:

$$r = \frac{q_{\min}}{q_{\max}} = \frac{T_{fs} - T_{fe}}{T_{ce} - T_{cs}} \qquad [4.43]$$

If the exchange is controlled by the cold fluid, then:

$$q_c > q_f \quad \Rightarrow \quad q_{\min} = q_f \quad ; \quad q_{\max} = q_c \qquad [4.44]$$

Noting that, in this case:

$$\Phi = q_c \left(T_{ce} - T_{cs}\right) = q_f \left(T_{fs} - T_{fe}\right) = q_{\max} \left(T_{ce} - T_{cs}\right) = q_{\min} \left(T_{fs} - T_{fe}\right) \qquad [4.45]$$

Therefore:

$$r = \frac{q_{\min}}{q_{\max}} = \frac{T_{ce} - T_{cs}}{T_{fs} - T_{fe}} \qquad [4.46]$$

4.2.3.4. *Efficiency of an exchanger*

The efficiency ε of an exchanger is defined by the ratio between the total flow exchanged (between hot and cold fluids), Φ and the "ideal" flow $\Phi_{\lim}$ that we would have if the exchanger were infinitely long:

$$\varepsilon = \frac{\Phi}{\Phi_{\lim}} \qquad [4.47]$$

In the case of an exchanger driven by hot fluid, ε is calculated by:

$$\Phi = q_c \left(T_{ce} - T_{cs}\right) = q_f \left(T_{fs} - T_{fe}\right) \qquad [4.48]$$

For the infinitely long exchanger $T_{cs} \rightarrow T_{fe}$:

$$\Phi_{\lim} = q_c \left(T_{ce} - T_{cs}\right) \rightarrow q_c \left(T_{ce} - T_{fe}\right) \qquad [4.49]$$

and

$$\varepsilon = \frac{\Phi}{\Phi_{\lim}} = \frac{q_c \left(T_{ce} - T_{cs}\right)}{q_c \left(T_{ce} - T_{fe}\right)} \qquad [4.50]$$

Therefore:

$$\varepsilon = \frac{\left(T_{ce} - T_{cs}\right)}{\left(T_{ce} - T_{fe}\right)} \qquad [4.51]$$

In the case of an exchanger driven by cold fluid, ε is calculated by:

$$\Phi = q_c \left(T_{ce} - T_{cs}\right) = q_f \left(T_{fs} - T_{fe}\right) \qquad [4.52]$$

For the infinitely long exchanger $T_{fs} \rightarrow T_{ce}$:

$$\Phi = q_f \left(T_{fs} - T_{fe}\right) \rightarrow q_f \left(T_{ce} - T_{fe}\right) \qquad [4.53]$$

and

$$\varepsilon = \frac{\Phi}{\Phi_{\lim}} = \frac{q_f \left(T_{fs} - T_{fe}\right)}{q_f \left(T_{ce} - T_{fe}\right)} \qquad [4.54]$$

Therefore: $\varepsilon = \dfrac{\left(T_{fs} - T_{fe}\right)}{\left(T_{ce} - T_{fe}\right)}$ [4.55]

4.2.3.5. *Number of transfer units*

We define the number of transfer units by:

$$N_{UT} = \frac{k\,S}{q} \quad [4.56]$$

In principle, there will be two numbers of transfer units to define: one on the hot side and one on the cold side:

$$N_{UTc} = \frac{k\,S}{q_c} \quad [4.57]$$

$$N_{UTf} = \frac{k\,S}{q_f} \quad [4.58]$$

As we will see later, in practice, only the N_{UT} defined from the minimum value of q will be taken into account. It is sometimes called $N_{UT\max}$.

Therefore:

$$N_{UT} = N_{UT\max} = \frac{k\,S}{q_{\min}} \equiv \max\left(\frac{k\,S}{q_{\min}}, \frac{k\,S}{q_{\max}}\right) \quad [4.59]$$

with: $q_{\min} = \min\left(q_c, q_f\right)$ [4.60]

In these conditions, the exchanged flow can be written as:

$$\Phi = q_{\min}\,\Delta T_{\max} \quad [4.61]$$

where: $\Delta T_{\max} = \max\left[\left(T_{ce} - T_{cs}\right), \left(T_{fs} - T_{fe}\right)\right]$ [4.62]

It can also be written as:

$$\Phi = k\,S\,\Delta T_{LM} \quad [4.63]$$

From these two writings, we deduce that:

$$\Phi = q_{\min} \, \Delta T_{\max} = k \, S \, \Delta T_{LM} \qquad [4.64]$$

Therefore: $N_{UT} = \; = \dfrac{k \, S}{q_{\min}} = \dfrac{\Delta T_{\max}}{\Delta T_{LM}}$ [4.65]

4.2.3.6. *Calculation of efficiencies*

With this in mind, the efficiencies can be related to the number of transfer units and the imbalance factor:

$$\varepsilon = \varepsilon\left(N_{UT}, r \right) \qquad [4.66]$$

The expression of ε will obviously depend on the type of exchanger used.

Calculation for a co-current exchanger

We will detail this calculation for the co-current exchanger.

We recall that:

$$\Delta T_{LM} = \frac{\left(T_{cs} - T_{fs}\right) - \left(T_{ce} - T_{fe}\right)}{Ln \dfrac{T_{cs} - T_{fs}}{T_{ce} - T_{fe}}} \qquad [4.67]$$

If the exchange is driven by the hot fluid, then:

$$q_c < q_f \quad \Rightarrow \quad q_{\min} = q_c \quad ; \quad q_{\max} = q_f \qquad [4.68]$$

$$r = \frac{q_{\min}}{q_{\max}} = \frac{T_{fs} - T_{fe}}{T_{ce} - T_{cs}} \qquad [4.69]$$

If the exchange is controlled by the cold fluid, then:

$$r = \frac{q_{\min}}{q_{\max}} = \frac{T_{ce} - T_{cs}}{T_{fs} - T_{fe}} \qquad [4.70]$$

In the case of an exchanger driven by hot fluid, ε is calculated by:

$$\varepsilon = \frac{\left(T_{ce} - T_{cs}\right)}{\left(T_{ce} - T_{fe}\right)} \quad [4.71]$$

In the case of an exchanger driven by cold fluid, ε is calculated by:

$$\varepsilon = \frac{\left(T_{fs} - T_{fe}\right)}{\left(T_{ce} - T_{fe}\right)} \quad [4.72]$$

We also recall that:

$$N_{UT} = \frac{k\,S}{q_{\min}} = \frac{\Delta T_{\max}}{\Delta T_{LM}} \quad [4.73]$$

Let us take the case of a co-current exchanger, where the hot fluid controls the transfer:

$$q_{\min} = q_c \quad [4.74]$$

$$N_{UT} = \frac{\Delta T_{\max}}{\Delta T_{LM}} = \frac{T_{ce} - T_{cs}}{\left(T_{ce} - T_{fe}\right) - \left(T_{cs} - T_{fs}\right)} Ln \frac{T_{ce} - T_{fe}}{T_{cs} - T_{fs}} \quad [4.75]$$

$$r = \frac{T_{fs} - T_{fe}}{T_{ce} - T_{cs}} \quad [4.76a]$$

$$\varepsilon = \frac{\left(T_{ce} - T_{cs}\right)}{\left(T_{ce} - T_{fe}\right)} \quad [4.76b]$$

With some rearrangement in the N_{UT}:

$$N_{UT} = \frac{T_{ce} - T_{cs}}{\left(T_{ce} - T_{fe}\right) - \left(T_{cs} - T_{fs}\right)} Ln \frac{T_{ce} - T_{fe}}{T_{cs} - T_{fs}} = \frac{T_{ce} - T_{cs}}{\left(T_{ce} - T_{cs}\right) + \left(T_{fs} - T_{fe}\right)} Ln \frac{T_{ce} - T_{fe}}{T_{cs} - T_{fs}} \quad [4.77]$$

$$= \frac{1}{1+\dfrac{T_{fs}-T_{fe}}{T_{ce}-T_{cs}}} Ln \frac{T_{ce}-T_{fe}}{T_{cs}-T_{fs}} = \frac{1}{1+r} Ln \frac{T_{ce}-T_{fe}}{T_{cs}-T_{fs}}$$

Simplifying the forward term of the logarithm by $T_{fs} - T_{fe}$.

Some manipulations in the logarithm argument give:

$$1-(1+r)\varepsilon = 1-\left(1+\frac{T_{fs}-T_{fe}}{T_{ce}-T_{cs}}\right)\frac{(T_{ce}-T_{cs})}{(T_{ce}-T_{fe})} = 1-\frac{T_{ce}-T_{cs}+T_{fs}-T_{fe}}{T_{ce}-T_{cs}}\frac{(T_{ce}-T_{cs})}{(T_{ce}-T_{fe})}$$

$$= \frac{(T_{ce}-T_{cs})-(T_{ce}-T_{cs}+T_{fs}-T_{fe})}{T_{ce}-T_{cs}}\frac{(T_{ce}-T_{cs})}{(T_{ce}-T_{fe})}$$

$$= \frac{(T_{fs}-T_{fe})}{(T_{ce}-T_{cs})}\frac{(T_{ce}-T_{cs})}{(T_{ce}-T_{fe})} = \frac{T_{fs}-T_{fe}}{T_{ce}-T_{fe}} = \frac{1}{\dfrac{T_{ce}-T_{fe}}{T_{cs}-T_{fs}}}$$

$$N_{UT} = \frac{1}{1+r} Ln \frac{T_{ce}-T_{fe}}{T_{cs}-T_{fs}} = \frac{-1}{1+r} Ln\left[1-(1+r)\varepsilon\right] \quad [4.78]$$

We can derive ε:

$$1-(1+r)\varepsilon = \exp\left(-(1+r)N_{UT}\right) \quad [4.79]$$

$$(1+r)\varepsilon = 1-\exp\left(-(1+r)N_{UT}\right)$$

The result is:

$$\varepsilon = \frac{1-\exp\left(-(1+r)N_{UT}\right)}{1+r} \quad [4.80]$$

Repeating the calculation for a transfer controlled by the cold fluid, we will explicitly find the same expression.

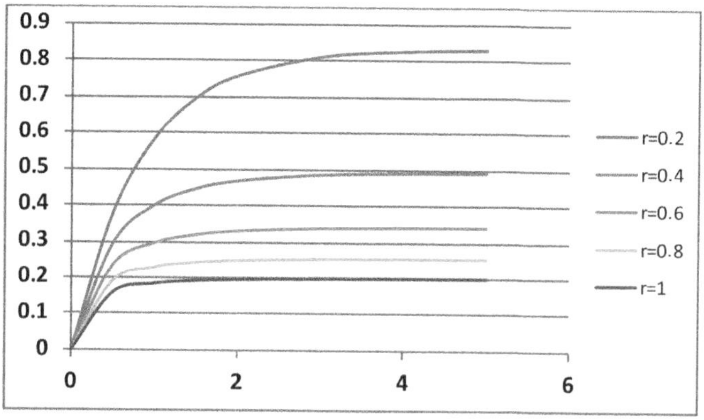

Figure 4.4. *Co-current exchanger. Efficiency according to N_{UT}. For a color version of this figure, see www.iste.co.uk/ledoux/heat3.zip*

Calculation for a countercurrent exchanger

A calculation like the one above gives the following expressions, whether the transfer is controlled by the hot or cold fluid:

$$N_{UT} = \frac{1}{r-1} Ln \frac{\varepsilon - 1}{r\varepsilon - 1} \quad [4.81]$$

$$\varepsilon = \frac{1-\exp\left(-(1-r)N_{UT}\right)}{1-r\exp\left(-(1-r)N_{UT}\right)} \quad [4.82]$$

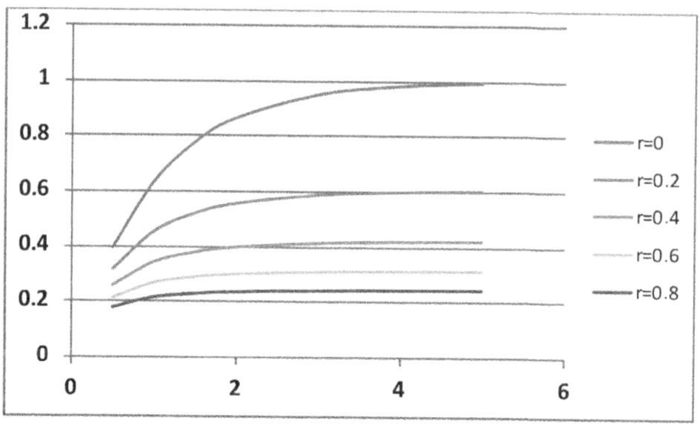

Figure 4.5. *Countercurrent exchanger. Efficiency according to N_{UT}. For a color version of this figure, see www.iste.co.uk/ledoux/heat3.zip*

Other geometries

We can establish such expressions between ε, r and N_{UT} for more complex exchanger geometries.

Hence, we are interested in cross-flow exchangers. In such exchangers, we distinguish between "stirred" flows, where the flow is not guided (flow around a bundle of tubes, for example), "non-stirred" flows (not excluding the fact that they are turbulent, in other words, strongly swirling), pipe flow, between plates or in rectangular cavities, etc.

The applicable results are summarized in Table 4.1.

Two co-current fluids	$\varepsilon = \dfrac{1-\exp(-(1+r)N_{UT})}{1+r}$
Two countercurrent fluids	$\varepsilon = \dfrac{1-\exp(-(1-r)N_{UT})}{1-r\exp(-(1-r)N_{UT})}$
Two non-stirred, cross-flow fluids	$\varepsilon = 1-\exp\dfrac{\exp\left(-r\,N_{UT}^{0,78}\right)-1}{r\,N_{UT}^{-0,22}}$
Two stirred fluids	$\varepsilon = \left[\dfrac{1}{1-\exp(-N_{UT})}+\dfrac{r}{1-\exp(-rN_{UT})}-\dfrac{1}{N_{UT}}\right]^{-1}$
Two cross-flow fluids, fluid controlling the non-stirred transfer	$N_{UT} = -Ln\left[1+\dfrac{1}{r}Ln\,(1-r\varepsilon)\right]$ $\varepsilon = \left(\dfrac{1}{r}\right)\left[1-\exp\left\{-r\left[1-\exp(-N_{UT})\right]\right\}\right]$
Two cross-flow fluids, fluid controlling the stirred transfer	$N_{UT} = -\dfrac{1}{r}Ln\,\left[1+r\,Ln\,(1-r\varepsilon)\right]$ $\varepsilon = 1-\exp\left\{-\dfrac{1}{r}\left[1-\exp(-r\,N_{UT})\right]\right\}$

Table 4.1. *Cross-flow exchangers*

NOTE.– Table 4.1 is by no means exhaustive. As the objective of this book is to combine basic culture and understanding, we will refer to more specialized works for a broader overview of these methods.

4.3. Conclusion

Note that the two methods presented here do not address the same problems. If we want to assess the performances of an existing exchanger, the *DTLM* method presents a challenge in that the performances of the exchanger are generally known (since they can also be searched for); the N_{UT} method will then be preferred. On the contrary, the *DTLM* method is adapted to the scaling of an exchanger with a given structure, since it allows for fixed performances to calculate an exchange surface.

The formulas given here will allow the reader to perform simple scaling on elementary exchangers. The details of the calculations given show that these methods are marked by a certain degree of approximation. The design of heat exchangers, especially in the early days of this technique, is strongly subject to experimental effort, an effort that the contributions of the digital world have undoubtedly reduced to exempt the designer.

4.4. An example of the application of the methods

As we will see, the application of these methods (*TLM*, N_{UT}) also calls upon the common sense of the user. There are obviously numerous application examples.

As mentioned in the introduction, this book is not a specific treatise on heat exchangers, and its size prevents it from being encyclopedic.

It appears more instructive, for us, to show the approach of the design of an elementary exchanger through an example.

EXAMPLE 4.1.– Laboratory benchtop exchanger

In a chemical research laboratory, we "tinker" with an exchanger designed to cool a liquid product.

Water is used as a cooling fluid. The product, being an aqueous solution, has the same properties as water. The exchanger is made up of three steel strips of length $L = 5\ m$ and width $l = 5\ cm$, welded $e = 5\,mm$ to one another, so as to form two channels of $50 * 5\,mm^2$ and of length $L = 10\,m$. The whole is immersed in a large volume of insulation.

NOTE.– From a practical point of view, the assembly must be rolled up, but the calculation will be made as if the exchanger were linear.

The flow rate of the product to be cooled is $q_P = 15\, l\,/\,mn$, the flow rate of the cooling water is $q_W = 20\, l\,/\,mn$.

The water enters the exchanger at T_{fe} = 15°C the product at T_{ce} = 50°C.

1) Is it better to use this exchanger in a co-current or countercurrent mode?

The edge effects and the influence of the regime establishments will be neglected.

2) We would like the product to leave the exchanger at T_{cs} = 40°C. **What should be the length** L **of the exchanger?**

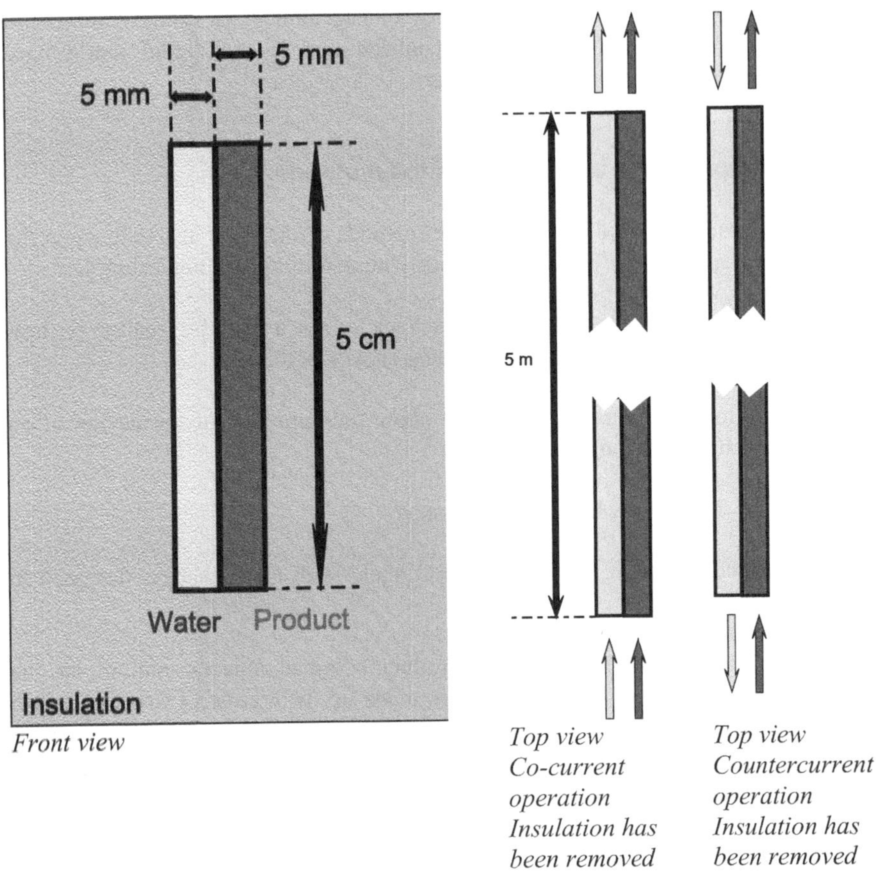

Figure 4.6. *The laboratory exchanger. For a color version of this figure, see www.iste.co.uk/ledoux/heat3.zip*

SOLUTION TO EXAMPLE 4.1.–

First, it is good practice to gather the information necessary to solve the problem.

Useful physical properties of water:

Density $\rho = 1000\ kg.m^{-3}$.

Dynamic viscosity $\mu = 1.10^{-3}\ Pl$.

Thermal conductivity $\lambda = 0.597\ Wm^{-1}K^{-1}$.

Mass heat capacity $c = 4182\ J\,kg^{-1}$.

From this, we can deduce:

Kinematic viscosity $\nu = \dfrac{\mu}{\rho} = \dfrac{1.10^{-3}}{1000} = 1.10^{-6}\ m^2 s^{-1}$.

Thermal diffusivity $a = \dfrac{\lambda}{\rho c_P} = \dfrac{0{,}597}{1000 * 4182} = 1.43.10^{-7}\ m^2\ s^{-1}$.

Prandtl number $\Pr = \dfrac{\nu}{a} = \dfrac{1.10^{-6}}{1.43.10^{-7}} = 6.99 \approx 7$.

Thermal conductivity of steel: $\lambda_{Fe} = 54\ W\ m^{-1}\ K^{-1}$.

SOLUTION.–

1) As a first step, we will calculate the constant *k* of the transfer resistance.

We will first think in terms of flow density. We will then apply the additivity of the transfer resistances.

As the whole is surrounded by insulation, we will assume that each channel can only exchange energy with its neighbor. The transfer will be done on a surface.

Conductive resistances

It is reduced to the steel plate, which is:

$$R_{thS} = \frac{e_{Fe}}{\lambda} = \frac{10^{-3}}{54} = 1.85.10^{-5}\ m^2\,K\,W^{-1} \quad [4.83]$$

Convection resistances

Let us calculate the Nusselt, then the convection coefficient for each of the branches.

We will take a flow between two plates as a model (the edge effects are neglected).

Let us determine the flow regimes.

Let us calculate the Reynolds.

The channel section is $s = 2.5.10^{-4}\ m^2$, the flow velocities are:

– for the product to be cooled, a volume flow rate of:

$$q_P = 15\ l.mn^{-1} = \frac{15.10^{-3}}{60} = 2.5.10^{-4}\ m^3\,s^{-1} \quad [4.84]$$

$$V_{qP} = \frac{q_P}{s} = \frac{2.5.10^{-4}}{2.5.10^{-4}} = 1\ m\,s^{-1} \quad [4.85]$$

– for cooling water, a volume flow rate of:

$$q_W = 20\ l.mn^{-1} = \frac{20.10^{-3}}{60} = 3.33.10^{-4}\ m^3\,s^{-1} \quad [4.86]$$

$$V_{qW} = \frac{q_P}{s} = \frac{3.33.10^{-4}}{2.5.10^{-4}} = 1.33\ m\,s^{-1} \quad [4.87]$$

Two Reynolds numbers will result from this. The thickness between two plates will be the reference length for the calculation of the Reynolds:

For the product:

$$R_{eP} = \frac{V_{qP}\, e}{\nu} = \frac{1*5.10^{-3}}{10^{-6}} = 5000 \quad [4.88]$$

For the water:

$$R_{eW} = \frac{V_{qW}\, e}{\nu} = \frac{1.33*5.10^{-3}}{10^{-6}} = 6650 \quad [4.89]$$

In both cases, the regime will be turbulent.

The Nusselt will therefore be calculated by the Colburn formula:

$$Nu = 0.023\, R_D^{0.8}\, \Pr^{1/3} \quad [4.90]$$

Therefore:

For the product:

$$Nu_{eP} = 0.023*5000^{0.8}*7^{1/3} = 40 \quad [4.91]$$

For the water:

$$Nu_{eW} = 0.023*6650^{0.8}*7^{1/3} = 50.3 \quad [4.92]$$

We deduce the convection coefficients:

$$h_P = \frac{\lambda Nu_{eP}}{e} = \frac{0.597*40}{5.10^{-3}} = 4776\ W\,m^{-2} \quad [4.93]$$

$$h_W = \frac{\lambda Nu_{eW}}{e} = \frac{0.597*50.3}{5.10^{-3}} = 6006\ W\,m^{-2} \quad [4.94]$$

Hence, the value of the thermal resistance R_{th}, at a point in the exchanger where the core temperatures are T_c and T_f:

$$R_{th} \text{ is defined by: } \varphi_W = \frac{T_c - T_f}{R_{th}} \quad [4.95]$$

and $R_{th} = R_{thS} + \dfrac{1}{h_W} + \dfrac{1}{h_P} = 1.85.10^{-5} + \dfrac{1}{4476} + \dfrac{1}{6006} \quad = 4.084.10^{-4}\ m^2\,K\,W^{-1}$ [4.96]

k is defined by:

$$d\,\Phi = k\left(T_c - T_f\right) d\,S \quad [4.97]$$

which is equivalent to writing:

$$\varphi_W = \frac{d\,\Phi}{d\,S} = k\left(T_c - T_f\right) \quad [4.98]$$

Therefore, by identification:

$$k = \frac{1}{R_{th}} = \frac{1}{4.084.10^{-4}} = 2448.5\ W\,m^{-2}\,K^{-1} \quad [4.99]$$

NOTE.– For a cylindrical geometry, care should be taken when defining the transfer resistances, especially if, as recommended in the parent work of this series on conduction, resistances per meter of tube are used.

We are looking for an exit temperature.

The N_{UT} method will be more suitable

$$q_C = q_P = \dot{m}_P\,c_P = \rho\,c\,q_{VP} = 1000 * 4182 * 2.5.10^{-4} = 1045.5\,J\,s^{-1} \quad [4.100]$$

$$q_f = q_W = \dot{m}_W\,c_W = \rho\,c\,q_{VW} = 1000 * 4182 * 3.33.10^{-4} = 1393\,J\,s^{-1} \quad [4.101]$$

$$q_C < q_f \quad [4.102]$$

The transfer is driven by the hot fluid.

We know that in this case, for an infinitely long exchanger, we would get:

$$T_{cs} \rightarrow T_{fe} \quad [4.103]$$

Therefore: $T_{Ps} \rightarrow T_{We}$ [4.104]

An imbalance factor $r = \frac{q_{\min}}{q_{\max}}$ is defined:

$$r = \frac{q_{\min}}{q_{\max}} = \frac{1045}{1393} = 0.75 \quad [4.105]$$

Let us also remember that:

$$r = \frac{q_{\min}}{q_{\max}} = \frac{T_{ce} - T_{cs}}{T_{fs} - T_{fe}} \quad [4.106]$$

The number of Transfer units, N_{UT} will be calculated from $q_{\min} = q_c$.

The transfer surface is:

$$S = L * l = 5 * 5.10^{-2} = 0.25\, m^2 \quad [4.107]$$

$$N_{UT} = \frac{k\,S}{q_{\min}} = \frac{k\,S}{q_c} = \frac{2448.5 * 0.25}{1045} = 0.585 \quad [4.108]$$

The efficiency ε of an exchanger is defined by the ratio between the total flow exchanged (between hot and cold fluids), Φ and the "ideal" flow $\Phi_{\lim}$ that we would have if the exchanger were infinitely long:

$$\varepsilon = \frac{\Phi}{\Phi_{\lim}} \quad [4.109]$$

In the case of an exchanger driven by the hot fluid, ε is calculated by:

$$\Phi = q_c\left(T_{ce} - T_{cs}\right) = q_f\left(T_{fs} - T_{fe}\right) \quad [4.110]$$

For the infinitely long exchanger $T_{cs} \rightarrow T_{fe}$:

$$\Phi_{\lim} = q_c\left(T_{ce} - T_{cs}\right) \rightarrow q_c\left(T_{ce} - T_{fe}\right) \quad [4.111]$$

Therefore:

$$\Phi_{\lim} = q_c\left(T_{ce} - T_{fe}\right) = 1045 * (50 - 15) = 36575\, W \quad [4.112]$$

Furthermore, the efficiency is calculated from the imbalance factor and the number of transfer units:$\varepsilon = \varepsilon(r, N_{UT})$.

Furthermore, we know that:

$$\Phi = q_c \left(T_{cs} - T_{ce}\right) \quad [4.113]$$

If we know ε, then we also know Φ. We can therefore determine the exit temperature of the hot fluid, in other words, the product to be cooled.

This expression will differ according to the operating mode of the exchanger, be it co-current or countercurrent.

If the exchanger operates in a co-current mode:

$$\varepsilon = \frac{1 - \exp\left(-\left(1 + r\right)N_{UT}\right)}{1 + r} \quad [4.114]$$

$$\varepsilon = \frac{1 - \exp\left(-\left(1 + 0.75\right) * 0.585\right)}{1 + 0.75} = 0.366 \quad [4.115]$$

If the exchanger operates in a countercurrent mode:

$$\varepsilon = \frac{1 - \exp\left(-\left(1 - r\right)N_{UT}\right)}{1 - r\exp\left(-\left(1 - r\right)N_{UT}\right)} = \frac{1 - \exp\left(-\left(1 - 0.75\right) * 0.585\right)}{1 - 0.75 * \exp\left(-\left(1 - 0.75\right) * 0.585\right)} = 0.386 \quad [4.116]$$

In co-current operation:

$$T_{cs} - T_{ce} = \frac{\Phi}{q_c} = \frac{\varepsilon\Phi_m}{q_c} = \frac{0.366 * 36575}{1045} = 12.8C \quad [4.117]$$

The product comes out of the exchanger at $50 - 17.4 = 37.1C$.

In a countercurrent operation:

$$T_{cs} - T_{ce} = \frac{\Phi}{q_c} = \frac{\varepsilon\Phi_m}{q_c} = \frac{0.386 * 36575}{1045} = 13.4C \quad [4.118]$$

The product comes out of the exchanger at $50-13.4=36.6C$.

A slight advantage to the countercurrent mode.

2) We would like to see the product temperature drop to $T_{cs}=40C$.

The exchange surface must be calculated, and the required channel length deduced.

To determine this, the DTLM method will be suitable to use

In this case, we do, in fact, know the exit and entry temperatures, as well as the exchange flow:

$$\Phi = q_c\left(T_{ce}-T_{cs}\right)=q_f\left(T_{fs}-T_{fe}\right) \quad [4.119]$$

with:

$$q_c=1045$$

$$q_f=1393$$

$$T_{fe}=15C$$

$$T_{ce}=50C$$

$$T_{cs}=40C \quad [4.120]$$

$$\frac{T_{fs}-T_{fe}}{T_{ce}-T_{cs}}=\frac{q_c}{q_f}=0.75$$

$$T_{fs}-T_{fe}=0.75\left(T_{ce}-T_{cs}\right)=0.75*10=7.5C$$

$$T_{fs}=T_{fe}+7.5=22.5C$$

And from this, the flow is deduced:

$$\Phi = q_c\left(T_{ce}-T_{cs}\right)=1045*20=20900W \quad [4.121]$$

The *DTLM* method tells us that:

$$\Phi = k\,S\,\Delta T_{LM} \quad [4.122]$$

with:

$$\Delta T_{LM} = \frac{\Delta T_1 - \Delta T_2}{Ln\dfrac{\Delta T_1}{\Delta T_2}} \quad [4.123]$$

If we operate in a co-current mode:

$$\Delta T_1 = T_{ce} - T_{fe} = 50 - 15 = 35\,C \quad [4.124]$$

$$\Delta T_2 = T_{cs} - T_{fs} = 40 - 22.5 = 17.5\,C \quad [4.125]$$

$$\Delta T_{LM} = \frac{\Delta T_1 - \Delta T_2}{Ln\dfrac{\Delta T_1}{\Delta T_2}} = \frac{35 - 17.5}{Ln\dfrac{35}{17.5}} = 25.24\,C \quad [4.126]$$

If we operate in a countercurrent mode:

$$\Delta T_1 = T_{ce} - T_{fs} = 50 - 22.5 = 27.5\,C \quad [4.127]$$

$$\Delta T_2 = T_{cs} - T_{fe} = 40 - 15 = 25\,C \quad [4.128]$$

$$\Delta T_{LM} = \frac{\Delta T_1 - \Delta T_2}{Ln\dfrac{\Delta T_1}{\Delta T_2}} = \frac{27.5 - 25}{Ln\dfrac{27.5}{25}} = 26.2\,C \quad [4.129]$$

with: $k = 2448.5$

And the flow to be exchanged:

$$\Phi = q_c\left(T_{ce} - T_{cs}\right) = 1045 * 10 = 10450\,W \quad [4.130]$$

The surface is determined by:

$$S = \frac{\Phi}{k\,\Delta T_{LM}} \quad [4.131]$$

The countercurrent mode appears to be the most advantageous:

$$\Delta T_{LM} = 26.2\,C \qquad [4.132]$$

And:

$$S = \frac{10450}{2448.5 * 26.2} = 0.16\,m^2 \qquad [4.133]$$

This would be a length of $3.25\,m$ for the exchanger.

In order of magnitude, this result is consistent with that of Question 1:

– the countercurrent mode is, once again, more efficient;

– a slightly longer exchanger ($L = 5m$) leads to a lower product exit temperature.

NOTE.– We can test the relation. Here, it is:

$$\Delta T_{LM} = \frac{\Delta T_1 + \Delta T_2}{2} \qquad [4.134]$$

Co-current:

$$\Delta T_{LM} = \frac{35 + 17.5}{2} = 26.25\,, \qquad [4.135]$$

to be compared with

$$\Delta T_{LM} = 25.24\,C\,. \qquad [4.136]$$

Countercurrent:

$$\Delta T_{LM} = \frac{27.5 + 25}{2} = 26.25\,, \qquad [4.137]$$

to be compared with

$$\Delta T_{LM} = 26.2\,C \qquad [4.138]$$

Appendices

Appendix 1

Physical Properties of Common Fluids

Temperature	Density	Mass heat capacity	Thermal conductivity	Dynamic viscosity	Kinematic viscosity	Prandtl number
T °C	ρ kg m^{-3}	c J kg^{-1}	λ W m^{-2} K^{-1}	μ Pl or Pa s^{-1}	ν m^2 s^{-1} or stokes (stk)	Pr
0	1,002	4,218	0.552	17.9	1.31	13.06
20	1,001	4,182	0.597	10.1	1.43	7.02
40	995	4,178	0.628	6.55	1.51	4.34
60	985	4,184	0.651	4.71	1.55	3.02
80	974	4,196	0.668	3.55	1.64	2.22
100	960	4,216	0.68	2.82	1.68	1.74
120	945	4,250	0.685	2.33	1.71	1.45
140	928	4,283	0.684	1.99	1.72	1.24
160	910	4,342	0.68	1.73	1.73	1.1
180	889	4,417	0.675	1.54	1.72	1
200	867	4,505	0.665	1.39	1.71	0.94
220	842	4,610	0.652	1.26	1.68	0.89
240	816	4,756	0.635	1.17	1.64	0.88
260	786	4,949	0.611	1.08	1.58	0.87
280	753	5,208	0.58	1.02	1.48	0.91
300	714	5,728	0.54	0.96	1.32	1.02

Table A1.1. *Physical properties of water*

Temperature	Density	Mass heat capacity	Thermal conductivity	Dynamic viscosity	Kinematic viscosity	Prandtl number
T °C	$\rho\ kg\ m^{-3}$	$c\ J\ kg^{-1}$	$\lambda\ W\ m^{-2}\ K^{-1}$	μ Pl or $Pa\ s^{-1}$	$\nu\ m^2\ s^{-1}$ or stokes (stk)	Pr
0	1.292	1,002	0.0242	1.72	1.86	0.72
20	1.204	1,006	0.0257	1.81	2.12	0.71
40	1.127	1,007	0.0272	1.9	2.4	0.7
60	1.059	1,008	0.0287	1.99	2.69	0.7
80	0.999	1,010	0.0302	2.09	3	0.7
100	0.946	1,012	0.0318	2.18	3.32	0.69
120	0.898	1,014	0.0333	2.27	3.66	0.69
140	0.854	1,016	0.0345	2.34	3.98	0.69
160	0.815	1,019	0.0359	2.42	4.32	0.69
180	0.779	1,022	0.0372	2.5	4.67	0.69
200	0.746	1,025	0.0386	2.57	5.05	0.68
220	0.700	1,028	0.0399	2.64	5.43	0.68
240	0.688	1,032	0.0412	2.72	5.8	0.68
260	0.662	1,036	0.0425	2.79	6.2	0.68
280	0.638	1,040	0.0437	2.86	6.59	0.68
300	0.616	1,045	0.0450	2.93	6.99	0.68

Table A1.2. *Physical properties of air*

Appendix 2

Physical Properties of Common Solids

	Density	Mass heat capacity	Thermal conductivity
	ρ $kg\,m^{-3}$	c $J\,kg^{-1}$	λ $W\,m^{-1}\,K^{-1}$
Carbon steel	7,833	465	54
Stainless steel 15% Cr, 10% Ni	7,864	460	20
Stainless steel 18% Cr, 8% Ni	7,816	460	16.3
Stainless steel 25% Cr, 20% Ni	7,864	460	13
Aluminum	2,707	896	204
Silver	10,525	234	407
Bronze 75% Cu, 25% Sn	8,800	377	188
Bronze 92% Cu, 8% AI	7,900	377	71
Carbon graphite	2,250	707	147
Chrome	2,118	7,160	449
Constantan 60% Cu, 40% Ni	8,922	410	22.7
Copper	8,954	383	386
Cupronickel 70% Cu, 30% Ni	8,900	377	29.3
Duralumin	2,787	3	164
Tin	7,304	226	64
Iron	7,870	452	73
Cast iron	7,849	460	59
Brass 70% Cu, 30% Zn	8,522	385	111
Magnesium	1,740	1,004	151
Gold	19,300	128	312
Platinum	21,400	140	69
Lead	11,373	130	35
Liquid sodium	930	1,381	84.5
Titanium	4,500	523	20.9
Tungsten	19,350	134	163
Zinc	7,144	384	112

Table A2.1. *Metallic materials: alloys*

	Density	Mass heat capacity	Thermal conductivity
	$\rho \; kg\,m^{-3}$	$c \; J\,kg^{-1}$	$\lambda \; W\,m^{-1}\,K^{-1}$
Asbestos	575	1,046	0.15
Asphalt	2,115	920	0.062
Rubber (natural)	1,150		0.28
Rubber (vulcanized)	1,100	2,010	0.13
Cardboard	86	2,030	0.048
Ice	920	2,040	1.88
Plexiglass	1,190	1,465	0.19
Porcelain	2,400	1,088	1.035
Polyethylene	929	1,830	0.46
PVC	1,459	930	0.21
Sand	1,515	800	0.2–1.0
Teflon	2,170	1,004	0.25
Wet earth	1,900	2,000	2
Dry earth	1,500	1,900	1
Glass	2,300	837	1.05
Pyrex glass	2,220	728	1.13

Table A2.2. *Miscellaneous materials*

	Density	Mass heat capacity	Thermal conductivity
	$\rho \; kg\,m^{-3}$	$c \; J\,kg^{-1}$	$\lambda \; W\,m^{-1}\,K^{-1}$
Slate	2,400	879	2.2
Basalt	2,850	881	1.6
Cavernous concrete	1,900	879	1.4
Solid concrete	2,300	878	1.75
Asphalt (cardboard)	1,050	1,305	0.23
Light hardwoods	525	3,143	0.15
Medium hardwoods	675	3,156	0.23
Very light hardwoods	375	3,147	0.12
Light softwoods	375	3,147	0.12
Medium–heavy softwoods	500	3,160	0.15
Very light softwoods	375	3,147	0.12
Clay brick	1,800	878	1.15
Hard limestone	2,450	882	2.4
Soft limestone	1,650	879	1
Tile	2,400	875	2.4

Okoume plywood	400	3,000	0.12
Pine plywood	500	3,000	0.15
Granite	2,600	881	3
Gravel (loose)	1,800	889	0.7
Sandstone	2,500	880	2.6
Lava	2,350	881	1.1
Marble	2,700	881	2.5
Gypsum	1,440	840	0.48
Shale	2,400	879	2.2

Table A2.3. *Materials used in construction*

	Density	**Mass heat capacity**	**Thermal conductivity**
	$\rho\ kg\,m^{-3}$	$c\ J\,kg^{-1}$	$\lambda\ W\,m^{-1}\,K^{-1}$
Balsa	140		0.054
Cotton	80	1,300	0.06
Rock wool (according to density)	20	880	0.047
	55	880	0.038
	135	880	0.041
Glass wool (according to density)	8	875	0.051
	10	880	0.045
	15	880	0.041
	40	880	0.035
Expanded cork	120	2,100	0.044
Carpet	200	1,300	0.06
Polyurethane foam (according to density)	32	1,300	0.03
	50	1,360	0.035
	85	1,300	0.045
PVC. rigid foam (according to density)	30	1,300	0.031
	40	1,300	0.041
Expanded polystyrene (according to density)	12	1,300	0.047
	14	1,300	0.043
	18	1,300	0.041

Table A2.4. *Insulation materials*

Appendix 3

Thermodynamic Properties of Water Vapor

For each temperature T of the vaporization stage, the table gives the saturating vapor tension p_{Sat}, the density of the liquid v_L, the density of the vapor v_V and the latent vaporization heat L.

T, °C	p_{Sat}, kPa	v_L, kg.m^{-3}	v_V, kg.m^{-3}	L, kJ.kg^{-1}
0.01	0.6113	0.001000	206.14	2501.3
5	0.8721	0.001000	147.12	2489.6
10	1.2276	0.001000	106.38	2477.7
15	1.7051	0.001001	77.93	2465.9
20	2.339	0.001002	57.79	2454.1
25	3.169	0.001003	43.36	2442.3
30	4.246	0.001004	32.89	2430.5
35	5.628	0.001006	25.22	2418.6
40	7.384	0.001008	19.52	2406.7
45	9.593	0.001010	15.26	2394.8
50	12.349	0.001012	12.03	2382.7
55	15.758	0.001015	9.568	2370.7
60	19.940	0.001017	7.671	2358.5
65	25.03	0.001020	6.197	2346.2

T, °C	p_{Sat}, MPa	v_L, $kg.m^{-3}$	v_V, $kg.m^{-3}$	L, J
70	31.19	0.001023	5.042	2333.8
75	38.58	0.001026	4.131	2321.4
80	47.39	0.001029	3.407	2308.8
85	57.83	0.001033	2.828	2296.0
90	70.14	0.001036	2.361	2283.2
95	84.55	0.001040	1.982	2270.2
100	0.10135	0.001044	1.6729	2257.0
105	0.12082	0.001048	1.4194	2243.7
110	0.14327	0.001052	1.2102	2230.2
115	0.16906	0.001056	1.0366	2216.5
120	0.19853	0.001060	0.8919	2202.6
125	0.2321	0.001065	0.7706	2188.5
130	0.2701	0.001070	0.6685	2174.2
135	0.3130	0.001075	0.5822	2159.6
140	0.3613	0.001080	0.5089	2144.7
145	0.4154	0.001085	0.4463	2129.6
150	0.4758	0.001091	0.3928	2114.3
155	0.5431	0.001096	0.3468	2098.6
160	0.6178	0.001102	0.3071	2082.6
165	0.7005	0.001108	0.2727	2066.2
170	0.7917	0.001114	0.2428	2049.5
175	0.8920	0.001121	0.2168	2032.4
180	1.0021	0.001127	0.19405	2015.0
185	1.1227	0.001134	0.17409	1997.1
190	1.2544	0.001141	0.15654	1978.8
195	1.3978	0.001149	0.14105	1960.0
200	1.5538	0.001157	0.12736	1940.7
205	1.7230	0.001164	0.11521	1921.0
210	1.9062	0.00117	0.10441	1900.7

$T, °C$	p_{Sat}, MPa	$v_L, kg.m^{-3}$	$v_V, kg.m^{-3}$	L, J
215	2.104	0.001181	0.09479	1879.9
220	2.318	0.001190	0.08619	1858.5
225	2.548	0.001199	0.07849	1836.5
230	2.795	0.001209	0.07158	1813.8
235	3.060	0.001219	0.06537	1790.5
240	3.344	0.001229	0.05976	1766.5
245	3.648	0.001240	0.05471	1741.7
250	3.973	0.001251	0.05013	1716.2
255	4.319	0.001263	0.04598	1689.8
260	4.688	0.001276	0.04221	1662.5
265	5.081	0.001289	0.03877	1634.4
270	5.499	0.001302	0.03564	1605.2
275	5.942	0.001317	0.03279	1574.9
280	6.412	0.001332	0.03017	1543.6
285	6.909	0.001348	0.02777	1511.0
290	7.436	0.001366	0.02557	1477.1
295	7.993	0.001384	0.02354	1441.8
300	8.581	0.001404	0.02167	1404.9
305	9.202	0.001425	0.019948	1366.4
310	9.856	0.001447	0.018350	1326.0
315	10.547	0.001472	0.016867	1283.5
320	11.274	0.001499	0.015488	1238.6
330	12.845	0.001561	0.012996	1140.6
340	14.586	0.001638	0.010797	1027.9
350	16.513	0.001740	0.008813	893.4
360	18.651	0.001893	0.006945	720.5
370	21.03	0.001213	0.004925	441.6
374.14	22.09	0.001155	0.003155	0

Appendix 4

The General Equations of Fluid Mechanics

A4.1. Reminders

A4.1.1. *Kinematic elements*

a) In the following writing in Cartesian coordinates, the following conventions will be used:

– The Cartesian coordinates axes, Ox, y, z, will be indexed as $i = 1.23$.

– We will use the Eulerian description of the flow.

The Eulerian components of the velocity $\vec{V}$ will be noted according to the requirements of the presentation $\vec{V}(u, v, w)$ or $\vec{V}(u_1, u_2, u_3)$.

b) The partial derivatives $\dfrac{\partial \vec{V}}{\partial t}$ must be distinguished from the total derivative $\dfrac{d\vec{V}}{dt}$ of the velocity, which constitutes the "true" acceleration of the fluid particle ($\dfrac{d\vec{V}}{dt}$ is still known as the hydrodynamic derivative or Lagrangian acceleration). By developing the expression of this acceleration, we can write:

$$\frac{d\vec{V}}{dt} = \frac{\partial \vec{V}}{\partial t} + \left(\vec{V}.\vec{grad}\right)\vec{V} \qquad [A4.1]$$

Or, projected onto the axis i:

$$\frac{du_i}{dt} = \frac{\partial u_i}{\partial t} + u_j . \frac{\partial u_i}{\partial x_j} \qquad \text{[A4.2]}$$

$\left(\vec{V}.\vec{grad}\right)\vec{V}$ is a special notation. We apply $\left(\vec{V}.\vec{grad}\right)$ successively to each component of $\vec{V}$. In this case, $\vec{V}.\vec{grad}\,u$, $\vec{V}.\vec{grad}\,v$ and $\vec{V}.\vec{grad}\,w$ are actually scalar products.

c) In what follows, we will use the Einstein notation in a rather systematic way.

Thus:

$$T_i = \sigma_{ij}\ n_j \text{ is equal to } T_i = \sum_{j=1}^{j=3} \sigma_{ij}\ n_j \qquad \text{[A4.3]}$$

The Kronecker symbol δ_{ij} is defined by:

$$\delta_{ij} = 1, \mathit{if}\ \ i = j \qquad \text{[A4.4]}$$

$$\delta_{ij} = 0, \mathit{if}\ \ i \neq j$$

A4.2. *Writing the principles*

The approach adopted for the writing of these principles consists of several steps:

1) We write a balanced equation for a mass of fluid identified in a domain, which we follow in its displacement. The writing in this case is Lagrangian.

2) We use a fundamental lemma to transform the relation into integrated expressions on a fixed domain (the initial domain observed at a given time).

3) We transform the possible surface integrals into volume integrals.

We reduce the set to a single volume integral which is zero; its integrand is then inevitably zero, and we deduce the so-called "local equation".

A4.2.1. Principle no. 1: Principle of continuity

This principle is covered in the chapter on Kinematics.

The temporal variation of the mass of fluid contained in a domain D, itself contained in a fixed surface S (with respect to the reference frame where the equations are written), is due to the flow of matter through S. So:

$$\frac{\partial}{\partial t}\iiint_D \rho \, d\omega = -\iint_S \rho \vec{V} \vec{n} \, dS \qquad [A4.5]$$

$$\frac{\partial}{\partial t}\iiint_D \rho \, d\omega = \iiint_D \frac{\partial \rho}{\partial t} d\omega$$

which turns into a local equation, the continuity equation:

By application of the divergence theorem:

$$-\iint_S \rho \vec{V} \vec{n} \, dS = -\iiint_D div \, \rho \vec{V} \, d\omega$$

The equation then becomes:

$$\iiint_D \frac{\partial \rho}{\partial t} + div \, \rho \vec{V} \, d\omega = 0 \qquad [A4.6]$$

Since the domain is arbitrary, the volume integral is only generally zero if the integrand is zero everywhere.

$$\frac{\partial \rho}{\partial t} + div \, \rho \vec{V} = \vec{0} \qquad [A4.7]$$

which can also be written as:

$$\frac{\partial \rho}{\partial t} + div \, \rho \vec{V} = \frac{\partial \rho}{\partial t} + \frac{\partial \rho u_i}{\partial x_i} = \frac{\partial \rho}{\partial t} + u_i \frac{\partial \rho}{\partial x_i} + \rho \frac{\partial u_i}{\partial x_i} = 0$$

Using the definition of the "hydrodynamic", "total" or "Lagrangian" derivative:

$$\frac{d}{dt} = \frac{\partial}{\partial t} + u_i \frac{\partial}{\partial x_i}$$

$$\frac{\partial \rho}{\partial t} + div\, \rho \vec{V} = \frac{\partial \rho}{\partial t} + \rho\, div \vec{V} = 0 \qquad \text{[A4.8]}$$

$\frac{d\rho}{dt}$ is the total derivative of the density. It expresses the variation of ρ following the fluid. [A4.7] allows us to show that the zero divergence is a necessary and sufficient condition for incompressibility: if ρ is constant, then $div\, \vec{V} = \vec{0}$; if $div\, \vec{V} = \vec{0}$, then $\frac{d\rho}{dt}$ is zero and the density is inevitably a constant in the flow.

A4.2.2. *Principle no. 2: The fundamental law of dynamics. The momentum equation*

Two theorems can be extracted from this principle: a local equation – the **momentum equation** and an **integrated form of the kinetic energy theorem**, which will be useful in order to establish the energy equation.

This law is written in the form of the center of mass theorem, for a mass of fluid contained in a domain *D*, bounded by a surface *S*, consisting of points that follow the fluid particles that are on *S*:

$$\frac{d}{dt} \iiint_D \rho \vec{V}\, d\omega = \iint_S \vec{F}_S\, dS + \iiint_D \rho \vec{F}_V\, d\omega \qquad \text{[A4.9]}$$

where $\vec{F}_S$ represents the surface forces (including pressure) and $\vec{F}_V$ represents the volume forces per unit mass.

This equation is modified, using a particular fundamental lemma:

$$\frac{d}{dt} \iiint_D \rho f\, d\omega = \iiint_D \rho \frac{df}{dt} d\omega$$

The equation is then rewritten in the form of integrals, all written on a fixed domain:

$$\iiint_D \rho \frac{d\vec{V}}{dt} d\omega = \iint_S \vec{F}_S \, dS + \iiint_D \rho \vec{F}_V \, d\omega \qquad [A4.10]$$

It is shown in class that the surface forces on a small area dS with a normal $\vec{n}\,(n_1, n_2, n_3)$, expressed from the resultant tension $\vec{T}$, can be expressed from the tensions on three surfaces normal to the three directions of the axes:

$$T_i = \sigma_{ij}\, n_j$$

This is also written in a somewhat hybrid form, $\vec{A}_S = \overline{\overline{T}}.\vec{n}$, which expresses that $\vec{F}_S$ is the contraction of the stress tensor and the unit vector of the normal. The explicit use of σ_{ij} seems clearer to us.

$$\sigma_{ij} - p\,\delta_{ij} + \tau_{ij}$$

δ_{ij} is the Kronecker symbol and p is defined by the opposite of the third of the stress tensor trace $p = -\frac{\sigma_{ii}}{3}$.

τ_{ij} can have, in the most common case, a very complex form (non-Newtonian fluids, with complex rheology). For the most common case of a Newtonian fluid, it is written as:

$$\tau_{ij} = \mu\left(\frac{\partial u_i}{\partial x_j} + \frac{\partial u_j}{\partial x_i}\right) + \eta \frac{\partial u_i}{\partial x_i} \qquad [A4.11]$$

μ is the dynamic viscosity and η is the volume viscosity. The volume viscosity term disappears for incompressible fluids.

Explaining the surface forces, we distinguish between pressure forces and forces resulting from rheological phenomena:

$$F_{Si} = \sigma_{ij}\, n_j = \left(-\frac{\partial\, p\delta_{ij}}{\partial x_j} + \tau_{ij} \right) n_j = -\frac{\partial p}{\partial x_i} + \tau_{ij}\, n_j \qquad \text{[A4.12]}$$

Throughout this passage, the transitions from one equality to the other are not immediate (they involve writing vector geometry formulas). The reader is encouraged to return to a complete course.

Returning to the equation of dynamics:

$$\iiint_D \rho \frac{d\vec{V}}{dt}\, d\omega = \iint_S \vec{F}_S\, dS + \iiint_D \rho\, \vec{F}_V\, d\omega$$

Projected on an axis in Cartesian coordinates, this equation becomes:

$$\iiint_D \rho \frac{du_i}{dt}\, d\omega = \iint_S F_{Si}\, dS + \iiint_D \rho\, F_{Vi}\, d\omega$$

Let us transform the double integral into a triple integral:

$$\iint_S F_{Si}\, dS = \iint_S \sigma_{ij} n_j\, dS$$

This integral can appear as the flow through S of a vector $<\sigma_i>$ of components $<\sigma_{ij}>$. This flow will then be equal to the volume integral of its divergence $\frac{\partial \sigma_{ij}}{\partial x_j}$:

$$\iint_S F_{Si}\, dS = \iint_S \sigma_{ij} n_j\, dS = \iiint_D \frac{\partial \sigma_{ij}}{\partial x_j}\, d\omega$$

which can be broken down into:

$$\iint_S F_{Si}\, dS = \iiint_D \frac{\partial \sigma_{ij}}{\partial x_j}\, d\omega = \iiint_D \frac{\partial\left(-p\delta_{ij} + \tau_{ij}\right)}{\partial x_j}\, d\omega = \iiint_D \frac{-\partial p}{\partial x_i} + \frac{\partial \tau_{ij}}{\partial x_j}\, d\omega$$

The equation can then be rewritten as:

$$\iiint_D \left[\rho \frac{d u_i}{d t} + \frac{\partial p}{\partial x_i} - \frac{\partial \tau_{ij}}{\partial x_j} - \rho F_{Vi} \right] d\omega = 0 \qquad [A4.13]$$

Since the domain is arbitrary, the volume integral is only generally zero if the integrand is zero everywhere:

$$\rho \frac{d u_i}{d t} + \frac{\partial p}{\partial x_i} - \frac{\partial \tau_{ij}}{\partial x_j} - \rho F_{Vi} = 0$$

This calculation leads to a set of local equations, commonly referred to as **the momentum equation**:

$$\rho \frac{d \vec{V}}{d t} = -\vec{grad}\, p + \vec{A}_S + \rho \vec{F}_V \qquad [A4.14]$$

The vector $\vec{A}_S$ integrates the surface forces other than pressure.

Therefore, the vector $\vec{A}_S$ will have the following as its ith component:

$$A_{Si} = \frac{\partial \tau_{ij}}{\partial x_j}$$

$\vec{F}_V$ integrates all of the volume forces per unit of mass.

These volume forces are often derived from a potential φ_V :

$$\vec{F}_V = -grad\, \varphi_V \qquad [A4.15]$$

Kinetic energy theorem

Let us use the dynamics equation to write the kinetic energy theorem for a fluid domain:

$$\rho \frac{d \vec{V}}{d t} = -\vec{grad}\, p + \vec{A}_S + \rho \vec{F}_V \qquad [A4.16]$$

with: $A_{Si} = \dfrac{\partial}{\partial x_j}\tau_{ij}$

By making the dot product of the two sides of the equation by the velocity:

$$\rho\frac{d\vec{V}}{dt}.\vec{V} = -\vec{grad}\, p.\vec{V} + \vec{A}_S.\vec{V} + \rho\vec{F}_V.\vec{V}$$

which can be rewritten as:

$$\rho\frac{du_i}{dt}.u_i = -u_i\frac{\partial p}{\partial x_i} + A_{Si}.u_i + \rho F_{Vi}.u_i$$

$$\rho\frac{du_i}{dt}.u_i = \rho\frac{d\,u_i.u_i}{2dt} = \rho\frac{de_c}{dt} \qquad [A4.17]$$

where $e_c = \dfrac{u_i.u_i}{2} = \dfrac{V^2}{2}$ is the local kinetic energy per unit of mass.

Integrated in the domain D, we obtain the following, noting that we systematically use the Einstein notation:

$$\iiint_D \rho\frac{de_c}{dt}d\omega = \iiint_D\left[-u_i\frac{\partial p}{\partial x_i} + \frac{\partial\tau_{ij}}{\partial x_j}.u_i + \rho F_{Vi}.u_i\right]d\omega \qquad [A4.18]$$

which we will rewrite as:

$$\iiint_D \rho\frac{de_c}{dt}d\omega = \iiint_D\left[-\frac{\partial pu_i}{\partial x_i} + p\frac{\partial u_i}{\partial x_i} + \frac{\partial\tau_{ij}.u_i}{\partial x_j} - \tau_{ij}.\frac{\partial u_i}{\partial x_j} + \rho F_{Vi}.u_i\right]d\omega \qquad [A4.19]$$

$$\iiint_D \rho\frac{de_c}{dt}d\omega = \iiint_D\left[-\frac{\partial pu_i}{\partial x_i} + p\,div\vec{V} + \frac{\partial\tau_{ij}.u_i}{\partial x_j} - \tau_{ij}.\frac{\partial u_i}{\partial x_j} + \rho F_{Vi}.u_i\right]d\omega \qquad [A4.20]$$

A4.2.3. *Principle no. 3: The first principle of thermodynamics. The energy equation*

We will write this principle for a domain *D*, following its displacement. The writing of the physical principle will be Lagrangian.

Exchanging work and heat with this field, we must take into account both the internal and kinetic energy of this element. Indeed, mechanical and "internal" energy are translated into an "average, directed" kinetic energy (one that can be said of the center of masses), and a kinetic energy "of the molecules in collision", determined by the speeds in relation to the center of masses of the molecules of the fluid particle. The exchange of energy will change for a small fluid particle.

Our balance will therefore show the kinetic energy of each fluid particle, which is added to the internal energy. The term "total" energy is then used. The fluid mechanic working on compressible flows is familiar with this formulation (see [LED 17b]). A reader who is used to classical thermodynamics problems involving "stationary" fluid volumes (fluid contained in a cylinder–piston system, for example) does not need to consider the "motion of the center of masses" of this fluid.

The first principle of thermodynamics is therefore written in a Lagrangian form as:

$$\frac{d}{dt}\iiint_D \rho(e_c + e_i)\,d\omega = \iint_S \vec{F}_S.\vec{V}.dS + \iiint_D \vec{F}_V.\vec{V}\,d\omega + \iint_S -\lambda\,\vec{Grad}\,T.\vec{n}\,dS + \iiint_D Q\,d\omega \qquad \text{[A4.21]}$$

where $e_c = \dfrac{V^2}{2}$ is the local internal energy per unit of mass; for a fluid, we can write:

$$de_i = c\,dT$$

where c is the mass heat capacity (c_V for a gas, c for a liquid).

Similarly, for the enthalpy per unit of mass of a gas:

$$dh_i = c_P\,dT$$

Q is the rate of mass heat creation in $J.m^{-3}$ (e.g. it may be radiation absorption in a gas).

We recall the fundamental lemma:

$$\frac{d}{dt}\iiint_D \rho f \, d\omega = \iiint_D \rho \frac{df}{dt} d\omega \qquad [A4.22]$$

We note that:

$$\iiint_D \rho \vec{F}_V.\vec{V}\, d\omega = \iiint_D \rho F_{Vi}\, u_i\, d\omega \qquad [A4.23]$$

Let us transform the energy equation into a sum of triple integrals.

$$\iiint_D \rho \frac{d(e_c + e_i)}{dt} d\omega = \iint_S \vec{F}_S.\vec{V}.dS + \iiint_D \vec{F}_V.\vec{V}\, d\omega + \iint_S -\lambda \vec{Grad}\, T.\vec{n}\, dS + \iiint_D Q\, d\omega \qquad [A4.24]$$

$$\vec{F}_S.\vec{V} = -p u_i n_i + \tau_{ij}\, u_i\, n_j$$

$$\iint_S \vec{F}_S.\vec{V}\, dS = \iint_S \left(-p u_i n_i + \tau_{ij}\, u_i\, n_j\right) dS = \iiint_D -\frac{\partial p u_i}{\partial x_i} d\omega + \iiint_D \frac{\partial \tau_{ij}\, u_i}{\partial x_j} d\omega \qquad [A4.25]$$

Indeed:

$\iint_S -p u_i n_i\, dS$ can be interpreted as the flow through S of a vector, whose i^{th} component is $p u_i$.

Therefore, by application of the divergence theorem:

$$\iint_S -p u_i n_i dS = \iiint_D \frac{\partial p u_i}{\partial x_i} d\omega \qquad [A4.26]$$

$\iint_S \tau_{ij}\, u_j\, n_i dS$ can be interpreted as the flow through S of a vector, whose i^{th} component is $\tau_{ij}\, u_j$.

Therefore, by application of the divergence theorem:

$$\iint_S -\tau_{ij}\, n_i\, dS = \iiint_D \frac{\partial\, \tau_{ij}\, n_j}{\partial\, x_i} d\omega \qquad \text{[A4.27]}$$

Noting that τ_{ij} is a symmetrical tensor.

Furthermore:

$$\iint_S -\lambda\, \vec{grad}\, T.\vec{n}\, dS = \iint_S -\lambda \frac{\partial T}{\partial\, x_i} n_i\, dS = \iiint_D -div\lambda\, \vec{grad}\, T\, d\omega \qquad \text{[A4.28]}$$

which, for the frequently encountered case where λ is constant, becomes:

$$\iint_S -\lambda\, \vec{grad}\, T.\vec{n}\, dS = \iint_S -\lambda \frac{\partial T}{\partial\, x_i} n_i\, dS = \iiint_D -\lambda\, \Delta T\, d\omega \qquad \text{[A4.29]}$$

Because: $div\, \vec{grad}\, T = \Delta T$

Ultimately, we find that:

$$\iiint_D \rho \frac{d\left(e_c + e_i \right)}{d\, t} d\omega \qquad \text{[A4.30]}$$

$$= \iiint_D -\frac{\partial\, pu_i}{\partial\, x_i} d\omega + \iiint_D \frac{\partial\, \tau_{ij}\, u_i}{\partial\, x_j} d\omega + \iiint_D \rho\, F_{Vi}\, u_i\, d\omega + \iiint_D -div\lambda\, \vec{grad}\, T\, d\omega + \iiint_D Q\, d\omega$$

Let us remove the kinetic energy theorem from this equation:

$$\iiint_D \rho \frac{d\, e_c}{d\, t} d\omega = \iiint_D \left[-\frac{\partial\, pu_i}{\partial\, x_i} + p\, div \vec{V} + \frac{\partial\, \tau_{ij} .u_i}{\partial\, x_j} - \tau_{ij} . \frac{\partial\, u_i}{\partial\, x_j} + \rho\, F_{Vi} .u_i \right] d\omega \qquad \text{[A4.31]}$$

Then comes the energy equation in integral form:

$$\iiint_D \rho \frac{d\,e_i}{d\,t} d\omega$$

$$= \iiint_D - p\,div\vec{V}\,d\omega + \iiint_D + \tau_{ij}.\frac{\partial\,u_i}{\partial\,x_j} d\omega + \iiint_D - div\lambda\,\vec{grad}\,T\,d\omega + \iiint_D Q\,d\omega \quad \text{[A4.32]}$$

By a process that is identical to the one used for the momentum equation, we establish the energy equation, a local equation:

$$\iiint_D \left[\rho \frac{d\,e_i}{d\,t} + p\,div\vec{V} - \tau_{ij}.\frac{\partial\,u_i}{\partial\,x_j} + div\lambda\,\vec{grad} - Q \right] d\omega = 0 \quad \text{[A4.33]}$$

Since the domain is arbitrary, the volume integral is generally only zero if the integrand is zero everywhere.

Ultimately, it comes to:

$$\rho \frac{d\,e_i}{d\,t} + p\,div\vec{V} = \tau_{ij}.\frac{\partial\,u_i}{\partial\,x_j} - div\lambda\,\vec{grad}\,T + Q \quad \text{[A4.34]}$$

We introduce the enthalpy per unit of mass h:

$$h = e_i + \frac{p}{\rho}$$

Because $\frac{1}{\rho}$ is the mass volume.

We get:

$$\frac{d\,h}{d\,t} = \frac{d\,e_i}{d\,t} + \frac{1}{\rho}\frac{d\,p}{d\,t} - \frac{p}{\rho^2}\frac{d\,\rho}{d\,t} = \frac{d\,e_i}{d\,t} + \frac{1}{\rho}\frac{d\,p}{d\,t} + \frac{p}{\rho^2}\left(\rho\,div\vec{V}\right) \quad \text{[A4.35]}$$

For the continuity equation, it is also written as: $\frac{d\,\rho}{d\,t} + \rho\,div\vec{V} = 0$

Therefore: $\rho\frac{d\,e_i}{d\,t}=\rho\frac{d\,h}{d\,t}-\frac{d\,p}{d\,t}-\ p\,div\vec{V}$ [A4.36]

$$\rho\frac{d\,e_i}{d\,t}+\ p\,div\vec{V}=\rho\frac{d\,h}{d\,t}-\frac{d\,p}{d\,t} \qquad [A4.37]$$

Hence, the energy equation, written in terms of enthalpy:

$$\rho\frac{d\,h}{d\,t}-\frac{d\,p}{d\,t}=\tau_{ij}.\frac{\partial\,u_i}{\partial\,x_j}-\ div\lambda\,\vec{g}rad\,T+Q \qquad [A4.38]$$

The variation of the local enthalpy thus varies according to the conduction of the various volumetric phenomena. The term $\tau_{ij}.\frac{\partial\,u_i}{\partial\,x_j}$ represents the contribution resulting from the thermalization of mechanical energy under the influence of rheology; more simply, by friction.

This term, which is generally quite complex, is called the dissipation function Φ:

$$\Phi=\tau_{ij}.\frac{\partial\,u_i}{\partial\,x_j}$$

$$\rho\frac{d\,h}{d\,t}-\frac{d\,p}{d\,t}=\Phi-\ div\lambda\,\vec{g}rad\,T+Q \qquad [A4.39]$$

As for a perfect gas, $dh=c_p\,dt$, the momentum equation can be written in the following form, which will often be used in this book:

$$\rho c_p\frac{d\,T}{d\,t}-\frac{d\,p}{d\,t}=-\ div\lambda\,\vec{g}rad\,T+Q+\Phi \qquad [A4.40]$$

A4.2.4. *Principle no. 4: Species continuity equation*

This fourth equation is, in fact, a system of continuity equations that are complementary to the general continuity equation. They are written when the fluid is a mixture of species. This equation then enters the field of thermal science in a flow where chemical reactions take place. This equation is also useful in process engineering, in the case where there is a simple diffusion without a reaction of the components of the mixture.

Let us look at a fluid density ρ, a mixture of n species, indexed as i.

The local composition of the mixture can be defined from different concentrations:

– **Molar concentration**: number of moles of the species i per unit volume in $mol.m^{-3}$.

– **Mole fraction**: ratio of the number of moles of a species i to the total number of moles, dimensionless.

– **Mass concentration** m_i: mass of the species per unit volume in $kg.m^{-3}$.

– **Mass fraction** c_i: ratio of the mass of the species i to the total mass per unit volume ρ, dimensionless $c_i = \frac{m_i}{\rho}$.

It is this last expression that we will use in the following.

We obviously have the following relation:

$$\sum_i c_i = \sum_i \frac{m_i}{\rho} = \frac{\sum_i m_i}{\rho} = \frac{\rho}{\rho} = 1 \qquad [A4.41]$$

Let us recall **Fick's law**, written in terms of mass fractions.

This law gives the mass flow q_{mi} of the species i in a mixture through a surface dS of a normal $\vec{n}$:

$$q_{mi} = -\rho\, D_i\, \vec{grad}\, c_i\, \vec{n}\, dS \qquad [A4.42]$$

The $-$ sign means that the flow is in the direction of decreasing concentrations.

Here, D_i is the diffusion coefficient of the species c_i in the mixture, in ms^{-2}. For Fick's molar-written law, this coefficient will obviously have a different value.

NOTE.– Fick's law is formally similar to Fourier's law. This is not accidental. Indeed, in Boltzmann's vision, diffusion is induced just as thermal conduction is to the effect of molecular shocks in a fluid. In many simple cases (two-dimensional boundary layer for example), this leads to a similarity of the heat and mass transport equations in a fluid. Noting that D_i has the same dimension as the kinematic viscosity and thermal diffusivity a, they are expressed in ms^{-2}. Their respective roles are, moreover, similar in the equations of momentum, energy and species continuity. We develop this analogy in Volume 4, Chapter 2.

Lastly, chemical reactions can occur between species. They are defined by the relations between the c_i. We will not be discussing the fact that they differ very strongly from one problem to another.

We will simply note that the reactions lead to appearances and disappearances, which in turn lead to terms of creation (positive) or disappearance (negative) of the species i, in other words, volume terms $Q_i\left(c_j\right)$. These terms, expressed in $kg.m^{-3}$, depend a priori on all or part of the c_j and kinetic constants of the reaction system that is specific to the problem studied.

Note also that chemistry involves:

$$\sum_i Q_i = 0 \qquad [A4.43]$$

Let us write the instantaneous mass balance for a species i in a fixed domain D:

$$\frac{\partial}{\partial t}\iiint_D \rho c_i \, d\omega = -\iint_S \rho c_i \vec{V}\vec{n}\, dS + \iint_S \rho D_i \, \overrightarrow{grad}\, c_i .\vec{n}\, dS + \iiint_D Q_i\left(c_j, k_j ...\right) d\omega \qquad [A4.44]$$

To obtain a local equation, we must reduce all integrals to a volume integral using the same process as before.

We have a negative sign in $-\iint_S \rho c_i \vec{V}\vec{n}\, dS$, because a flow directed in the direction of the normal, which is oriented outwards, constitutes a mass loss in the domain D.

We have a positive sign in $\iint_S \rho D_i \, \overrightarrow{grad}\, c_i .\vec{n}\, dS$, because the diffusive flow is oriented in the direction of $-\overrightarrow{grad}\, c_i$, and if it is directed in the direction of the normal, it constitutes a mass loss in the domain D.

We will get:

$$-\iint_S \rho c_i \vec{V}\vec{n}\, dS = \iiint_D div\ \rho c_i \vec{V}\ d\omega \quad [A4.44]$$

$$\iint_S \rho D_i \overrightarrow{grad}\, c_i .\vec{n}\, dS = \iiint_D div\, \rho D_i \overrightarrow{grad}\, c_i\ d\omega \quad [A4.45]$$

The sum of these two terms gives:

$$\begin{aligned}&-\iiint_D div\ \rho c_i \vec{V}\ d\omega + \iiint_D div\, \rho D_i \overrightarrow{grad}\, c_i\ d\omega\\ &= \iiint_D -c_i\ div\, \rho \vec{V} - \rho \vec{V} div\, c_i + div\, \rho D_i \overrightarrow{grad}\, c_i\, d\omega\end{aligned} \quad [A4.46]$$

The continuity equation of the species i is rewritten as:

$$\begin{aligned}&\frac{\partial}{\partial t}\iiint_D \rho\, c_i\, d\omega = \iiint_D -c_i\ div\, \rho \vec{V} + -\rho \vec{V} div\, c_i + div\, \rho D_i \overrightarrow{grad}\, c_i\, d\omega\\ &+\iiint_D Q_i\left(c_j, k_j ...\right) d\omega\end{aligned} \quad [A4.47]$$

Or as:

$$\frac{\partial}{\partial t}\iiint_D \rho\, c_i\, d\omega = \iiint_D \frac{\partial}{\partial t}\rho\, c_i\, d\omega$$

$$\iiint_D \left[\frac{\partial}{\partial t}\rho\, c_i + c_i\ div\, \rho \vec{V} + \rho \vec{V} div\, c_i - div\, \rho D_i \overrightarrow{grad}\, c_i - Q_i\left(c_j, k_j ...\right)\right] d\omega = 0 \quad [A4.48]$$

The domain being unspecified, the triple integral can only be zero if the integrand is zero everywhere:

$$\frac{\partial}{\partial t}\rho c_i + c_i\ div\, \rho \vec{V} + \rho \vec{V} div\, c_i - div\, \rho D_i \overrightarrow{grad}\, c_i - Q_i\left(c_j, k_j ...\right) = 0 \quad [A4.49]$$

We can rearrange the terms:

$$\frac{\partial}{\partial t}\rho c_i + c_i\, div\, \rho\vec{V} + \rho\vec{V} div\, c_i = \rho\frac{\partial c_i}{\partial t} + \rho\vec{V} div\, c_i + c_i\frac{\partial \rho}{\partial t} + c_i\, div\, \rho\vec{V} \quad [A4.50]$$

We know that:

$$\rho\frac{\partial}{\partial t} + \rho\vec{V} div = \rho\frac{d}{dt} \quad [A4.51]$$

Therefore:

$$\rho\frac{\partial c_i}{\partial t} + \rho\vec{V} div\, c_i + c_i\frac{\partial \rho}{\partial t} + c_i\, div\, \rho\vec{V} = \rho\frac{dc_i}{dt} + c_i\frac{d\rho}{dt} = \rho\frac{dc_i}{dt} \quad [A4.52]$$

Because according to the continuity equation of the mixed fluid: $\frac{d\rho}{dt} = 0$

We rewrite and rearrange:

$$\rho\frac{d c_i}{dt} - div\, \rho D_i\, \overrightarrow{grad}\, c_i - Q_i\left(c_j, k_j ...\right) = 0 \quad [A4.53]$$

Ultimately, considering in the most frequent case that the diffusion coefficient is independent of its concentration (which is not always true):

$$\rho\frac{d c_i}{dt} = D_i\, div\, \rho\, \overrightarrow{grad}\, c_i + Q_i\left(c_j, k_j ...\right) \quad [A4.54]$$

NOTE.–

We thus have n independent equations.

The general continuity equation is not an $(n+1)^{th}$ equation.

Indeed, by adding the n equations for each i:

$$\sum_i \rho\frac{d c_i}{dt} = \sum_i -div\, \rho c_i \vec{V} + \sum_i Q_i \quad [A4.55]$$

The conservation of matter leads to the following relation:

$$\sum_i Q_i = 0$$

$$\sum_i \rho \frac{d c_i}{d t} = \sum_i -div\, \rho c_i \vec{V} \qquad [A4.56]$$

$$\sum_i \frac{d \rho c_i}{d t} - \sum_i \rho \frac{d c_i}{d t} = \sum_i -div\, \rho c_i \vec{V} \qquad [A4.57]$$

We thus find the continuity equation.

A4.3. The equations: developed writing and simplifications

A4.3.1. *The complete equations of fluid mechanics, known as "local equations", will be written as follows, in Cartesian coordinates*

Continuity equation

$$\frac{\partial \rho}{\partial t} + \frac{\partial \rho u}{\partial x} + \frac{\partial \rho v}{\partial y} + \frac{\partial \rho w}{\partial z} = 0 \qquad [A4.58]$$

Momentum equation, projected on the three axes

Projection on Ox:

$$\rho\left[\frac{\partial u}{\partial t} + u\frac{\partial u}{\partial x} + v\frac{\partial u}{\partial y} + w\frac{\partial u}{\partial z}\right] = -\frac{\partial p}{\partial x} + \frac{\partial}{\partial x}\left[2\mu\left(\frac{\partial u}{\partial x}\right)\right]$$

$$+\frac{\partial}{\partial y}\left[\mu\left(\frac{\partial u}{\partial y} + \frac{\partial v}{\partial x}\right)\right] + \frac{\partial}{\partial z}\left[\mu\left(\frac{\partial u}{\partial z} + \frac{\partial w}{\partial x}\right)\right] + \frac{\partial}{\partial x}\left[\eta\left(\frac{\partial u}{\partial x} + \frac{\partial v}{\partial y} + \frac{\partial w}{\partial z}\right)\right] + \rho F_{Vx} \qquad [A4.59]$$

Projection on Oy:

$$\rho\frac{\partial v}{\partial t} + u\frac{\partial v}{\partial x} + v\frac{\partial v}{\partial y} + w\frac{\partial v}{\partial z} = -\frac{\partial p}{\partial y} + \frac{\partial}{\partial x}\left[\mu\left(\frac{\partial u}{\partial y} + \frac{\partial v}{\partial x}\right)\right]$$

$$+\frac{\partial}{\partial y}\left[2\mu\left(\frac{\partial u}{\partial y}\right)\right]+\frac{\partial}{\partial z}\left[\mu\left(\frac{\partial v}{\partial z}+\frac{\partial w}{\partial y}\right)\right]+\frac{\partial}{\partial x}\left[\eta\left(\frac{\partial u}{\partial x}+\frac{\partial v}{\partial y}+\frac{\partial w}{\partial z}\right)\right]+\rho F_{Vy} \qquad [A4.60]$$

Projection on Oz:

$$\rho\frac{\partial w}{\partial t}+u\frac{\partial w}{\partial x}+v\frac{\partial u}{\partial y}+w\frac{\partial w}{\partial z}=-\frac{\partial p}{\partial z}+\frac{\partial}{\partial x}\left[\mu\left(\frac{\partial u}{\partial z}+\frac{\partial w}{\partial x}\right)\right]$$

$$+\frac{\partial}{\partial y}\left[\mu\left(\frac{\partial w}{\partial y}+\frac{\partial v}{\partial z}\right)\right]+\frac{\partial}{\partial z}\left[2\mu\left(\frac{\partial w}{\partial z}\right)\right]+\frac{\partial}{\partial x}\left[\eta\left(\frac{\partial u}{\partial x}+\frac{\partial v}{\partial y}+\frac{\partial w}{\partial z}\right)\right]+\rho F_{Vz} \qquad [A4.61]$$

Energy equation (for a perfect gas)

$$\rho c_p\frac{dT}{dt}-\frac{dp}{dt}=-\ div\lambda\,\vec{grad}\,T+Q+\Phi \qquad [A4.62]$$

$$\rho c_p\left(\frac{\partial T}{\partial t}+u\frac{\partial T}{\partial x}+v\frac{\partial T}{\partial y}+w\frac{\partial T}{\partial z}\right)-\left(\frac{\partial p}{\partial t}+u\frac{\partial p}{\partial x}+v\frac{\partial p}{\partial y}+w\frac{\partial p}{\partial z}\right)$$
$$=-\left(\frac{\partial}{\partial x}\lambda\frac{\partial T}{\partial x}+\frac{\partial}{\partial y}\lambda\frac{\partial T}{\partial y}+\frac{\partial}{\partial z}\lambda\frac{\partial T}{\partial z}\right)+Q+\Phi \qquad [A4.63]$$

Φ has a complex form. For a Newtonian fluid, it will be written as:

The notation used here is:

$\tau_{xy}=\tau_{12}$, etc.

We will thus have:

$$\tau_{xy}=\mu\left(\frac{\partial u}{\partial y}+\frac{\partial v}{\partial x}\right),\ \text{etc.}$$

$$\Phi=\mu\left[\left(\frac{2\partial u}{\partial y}\right)\frac{\partial u}{\partial x}+\left(\frac{\partial u}{\partial y}+\frac{\partial v}{\partial x}\right)\frac{\partial u}{\partial y}+\left(\frac{\partial u}{\partial z}+\frac{\partial w}{\partial x}\right)\frac{\partial u}{\partial z}\right]$$

$$+\mu\left[\left(\frac{\partial v}{\partial x}+\frac{\partial u}{\partial y}\right)\frac{\partial v}{\partial x}+\left(\frac{2\partial v}{\partial y}\right)\frac{\partial v}{\partial y}+\left(\frac{\partial v}{\partial z}+\frac{\partial w}{\partial y}\right)\frac{\partial v}{\partial z}\right] \quad [A4.64]$$

$$+\mu\left[\left(\frac{\partial w}{\partial x}+\frac{\partial u}{\partial z}\right)\frac{\partial w}{\partial x}+\left(\frac{\partial w}{\partial y}+\frac{\partial v}{\partial z}\right)\frac{\partial w}{\partial y}+\left(\frac{2\partial w}{\partial z}\right)\frac{\partial w}{\partial z}\right]$$

Species continuity equation

$$\rho\frac{d\,c_i}{d\,t}=D_i\,div\,\rho\,\overrightarrow{grad}\,c_i+Q_i\left(c_j,k_j...\right) \quad [A4.65]$$

This is written in expanded form:

$$\rho\left(\frac{\partial c_i}{\partial t}+u\frac{\partial c_i}{\partial x}+v\frac{\partial c_i}{\partial y}+w\frac{\partial c_i}{\partial z}\right)=-D_i\left(\frac{\partial}{\partial x}\rho\frac{\partial T}{\partial x}+\frac{\partial}{\partial y}\rho\frac{\partial T}{\partial y}+\frac{\partial}{\partial z}\rho\frac{\partial T}{\partial z}\right)+Q_i \quad [A4.66]$$

A4.3.2. *Application of simplifying assumptions*

There is no general solution to these equations. Any fluid mechanics problem will involve the establishment of more or less restrictive assumptions in order to simplify the equations.

We can try to get rid of such approximations by using a direct numerical simulation (DNS), which is very demanding in terms of computing resources and time.

The flow is often two-dimensional

This is a practical application case. The velocity only has two components and the momentum equation only has two projections, on *Ox* and *Oy*.

Continuity equation

$$\frac{\partial\rho}{\partial t}+\frac{\partial\rho u}{\partial x}+\frac{\partial\rho v}{\partial y}=0 \quad [A4.67]$$

Momentum equation

$$\rho\left[\frac{\partial u}{\partial t}+u\frac{\partial u}{\partial x}+v\frac{\partial u}{\partial y}\right]$$

$$=-\frac{\partial p}{\partial x}+\frac{\partial}{\partial x}\left[2\mu\left(\frac{\partial u}{\partial x}\right)\right]+\frac{\partial}{\partial y}\left[\mu\left(\frac{\partial u}{\partial y}+\frac{\partial v}{\partial x}\right)\right]+\frac{\partial}{\partial x}\left[\eta\left(\frac{\partial u}{\partial x}+\frac{\partial v}{\partial y}\right)\right]+\rho F_{Vx} \quad \text{[A4.68]}$$

$$\rho\frac{\partial v}{\partial t}+u\frac{\partial v}{\partial x}+v\frac{\partial v}{\partial y}$$

$$=-\frac{\partial p}{\partial y}+\frac{\partial}{\partial x}\left[\mu\left(\frac{\partial u}{\partial y}+\frac{\partial v}{\partial x}\right)\right]+\frac{\partial}{\partial y}\left[2\mu\left(\frac{\partial u}{\partial y}\right)\right]++\frac{\partial}{\partial x}\left[\eta\left(\frac{\partial u}{\partial x}+\frac{\partial v}{\partial y}\right)\right] \quad \text{[A4.69]}$$

$$+\rho F_{Vy}$$

Energy equation

$$\rho c_p\left(\frac{\partial T}{\partial t}+u\frac{\partial T}{\partial x}+v\frac{\partial T}{\partial y}\right)-\left(\frac{\partial p}{\partial t}+u\frac{\partial p}{\partial x}+v\frac{\partial p}{\partial y}\right) \quad \text{[A4.70]}$$

$$=-\left(\frac{\partial}{\partial x}\lambda\frac{\partial T}{\partial x}+\frac{\partial}{\partial y}\lambda\frac{\partial T}{\partial y}\right)+Q+\Phi$$

$$\Phi=\mu\left[\left(\frac{2\partial u}{\partial y}\right)\frac{\partial u}{\partial x}+\left(\frac{\partial u}{\partial y}+\frac{\partial v}{\partial x}\right)\frac{\partial u}{\partial y}+\left(\frac{\partial v}{\partial x}+\frac{\partial u}{\partial y}\right)\frac{\partial v}{\partial x}+\left(\frac{2\partial v}{\partial y}\right)\frac{\partial v}{\partial y}\right] \quad \text{[A4.71]}$$

Species continuity equation

$$\rho\left(\frac{\partial c_i}{\partial t}+u\frac{\partial c_i}{\partial x}+v\frac{\partial c_i}{\partial y}\right)=-D_i\left(\frac{\partial}{\partial x}\rho\frac{\partial T}{\partial x}+\frac{\partial}{\partial y}\rho\frac{\partial T}{\partial y}\right)+Q_i \quad \text{[A4.72]}$$

The flow can be incompressible

For an incompressible fluid, the equations simplify to:

Continuity equation

$$div\vec{V} = 0 \qquad [A4.73]$$

Dynamics equation

$$\rho\frac{d\vec{V}}{dt} = -\vec{grad}\, p + \vec{T} \qquad [A4.74]$$

When all physical properties of the fluid are constant (including incompressibility), and the fluid is Newtonian we obtain the so-called **Navier–Stokes equations**. In writing these equations, the introduction of the kinematic viscosity, resulting from a division by ρ, is necessary.

The Navier–Stokes equations are written as follows, for a two-dimensional flow:

$$\frac{\partial u}{\partial x} + \frac{\partial v}{\partial y} = 0 \qquad [A4.75]$$

$$\frac{\partial u}{\partial t} + u\frac{\partial u}{\partial x} + v\frac{\partial u}{\partial y} = -\frac{1}{\rho}\frac{\partial p}{\partial x} + 2\nu\frac{\partial^2 u}{\partial x^2} + \nu\left[\frac{\partial^2 u}{\partial y^2} + \frac{\partial v}{\partial x \partial y}\right] + \rho F_{Vx} \qquad [A4.76]$$

$$\frac{\partial v}{\partial t} + u\frac{\partial v}{\partial x} + v\frac{\partial v}{\partial y} = -\frac{1}{\rho}\frac{\partial p}{\partial y} + \nu\left(\frac{\partial^2 u}{\partial x \partial y} + \frac{\partial^2 v}{\partial x^2}\right) + \frac{\partial}{\partial y}2\nu\left(\frac{\partial^2 u}{\partial y^2}\right) + \rho F_{Vy} \qquad [A4.77]$$

where ν is the kinematic viscosity (expressed in myriastokes, or more usually in $m^2.s^{-1}$ in the SI system):

$$\nu = \frac{\mu}{\rho}$$

Given that the divergence of the velocity is null for an incompressible fluid, it will be noted that the term containing the “volume viscosity” η disappears. This is a general property for all incompressible fluids, whether they are Newtonian or not.

The energy equation is also simplified

We introduce the thermal diffusivity $a = \dfrac{\lambda}{\rho c_p}$

$$\left(\frac{\partial T}{\partial t}+u\frac{\partial T}{\partial x}+v\frac{\partial T}{\partial y}\right)-\frac{1}{\rho c_p}\left(\frac{\partial p}{\partial t}+u\frac{\partial p}{\partial x}+v\frac{\partial p}{\partial y}\right) \qquad \text{[A4.78]}$$

$$=- a\left(\frac{\partial^2 T}{\partial x^2}+\frac{\partial^2 T}{\partial y^2}\right)+Q+\frac{v}{c_p}$$

$$\left[\left(\frac{2\partial u}{\partial y}\right)\frac{\partial u}{\partial x}+\left(\frac{\partial u}{\partial y}+\frac{\partial v}{\partial x}\right)\frac{\partial u}{\partial y}+\left(\frac{\partial v}{\partial x}+\frac{\partial u}{\partial y}\right)\frac{\partial v}{\partial x}+\left(\frac{2\partial v}{\partial y}\right)\frac{\partial v}{\partial y}\right]$$

The species continuity equation i simplifies to:

$$\frac{\partial c_i}{\partial t}+u\frac{\partial c_i}{\partial x}+v\frac{\partial c_i}{\partial y}=- D_i\left(\frac{\partial^2 T}{\partial x^2}+\frac{\partial^2 T}{\partial y^2}\right)+Q_i \qquad \text{[A4.79]}$$

Finally, **in permanent flows**, the time derivatives disappear and the systems become:

Continuity equation

$$\frac{\partial u}{\partial x}+\frac{\partial v}{\partial y}=0 \qquad \text{[A4.80]}$$

Momentum equations

$$u\frac{\partial u}{\partial x}+v\frac{\partial u}{\partial y} =-\frac{1}{\rho}\frac{\partial p}{\partial x}+2v\frac{\partial^2 u}{\partial x^2}+v\left[\frac{\partial^2 u}{\partial y^2}+\frac{\partial v}{\partial x\partial y}\right]+\rho F_{Vx} \qquad \text{[A4.81]}$$

$$u\frac{\partial v}{\partial x}+v\frac{\partial v}{\partial y} =-\frac{1}{\rho}\frac{\partial p}{\partial y}+v\left(\frac{\partial^2 u}{\partial x\partial y}+\frac{\partial^2 v}{\partial x^2}\right)+\frac{\partial}{\partial y}2v\left(\frac{\partial^2 u}{\partial y^2}\right)+\rho F_{Vy} \qquad \text{[A4.82]}$$

Energy equation

$$\left(u\frac{\partial T}{\partial x}+v\frac{\partial T}{\partial y}\right)-\frac{1}{\rho c_p}\left(\frac{\partial p}{\partial t}+u\frac{\partial p}{\partial x}+v\frac{\partial p}{\partial y}\right) \qquad \text{[A4.83]}$$

$$= - a\left(\frac{\partial^2 T}{\partial x^2} + \frac{\partial^2 T}{\partial y^2}\right) + Q + \frac{\nu}{c_p}$$

$$\left[\left(\frac{2\partial u}{\partial y}\right)\frac{\partial u}{\partial x} + \left(\frac{\partial u}{\partial y} + \frac{\partial v}{\partial x}\right)\frac{\partial u}{\partial y} + \left(\frac{\partial v}{\partial x} + \frac{\partial u}{\partial y}\right)\frac{\partial v}{\partial x} + \left(\frac{2\partial v}{\partial y}\right)\frac{\partial v}{\partial y}\right]$$

Species continuity equation i

$$u\frac{\partial c_i}{\partial x} + v\frac{\partial c_i}{\partial y} = - D_i\left(\frac{\partial^2 T}{\partial x^2} + \frac{\partial^2 T}{\partial y^2}\right) + Q_i \qquad [A4.84]$$

Appendix 5

The Dynamic and Thermal Laminar Boundary Layer

Equations and analytical solutions

A5.1. Establishment of equations: boundary layer simplifications

We will place ourselves here in the case of permanent, two-dimensional flows, with constant physical properties (and thus, incompressible).

A5.1.1. *The equations have been established in Appendix 4*

Continuity equation

$$\frac{\partial u}{\partial x}+\frac{\partial v}{\partial y}=0 \qquad [A5.1]$$

Momentum equations

$$u\frac{\partial u}{\partial x}+v\frac{\partial u}{\partial y}=-\frac{1}{\rho}\frac{\partial p}{\partial x}+2\nu\frac{\partial^2 u}{\partial x^2}+\nu\left[\frac{\partial^2 u}{\partial y^2}+\frac{\partial v}{\partial x\partial y}\right]+\rho F_{Vx} \qquad [A5.2]$$

$$u\frac{\partial v}{\partial x}+v\frac{\partial v}{\partial y}=-\frac{1}{\rho}\frac{\partial p}{\partial y}+\nu\left(\frac{\partial^2 u}{\partial x\partial y}+\frac{\partial^2 v}{\partial x^2}\right)+\frac{\partial}{\partial y}2\nu\left(\frac{\partial^2 u}{\partial y^2}\right)+\rho F_{Vy} \qquad [A5.3]$$

Energy equation

$$\left(u\frac{\partial T}{\partial x}+v\frac{\partial T}{\partial y}\right)-\frac{1}{\rho c_p}\left(\frac{\partial p}{\partial t}+u\frac{\partial p}{\partial x}+v\frac{\partial p}{\partial y}\right) \quad [A5.4]$$

$$=-\ a\left(\frac{\partial^2 T}{\partial x^2}+\frac{\partial^2 T}{\partial y^2}\right)+Q+\frac{v}{c_p}$$

$$\left[\left(\frac{2\partial u}{\partial y}\right)\frac{\partial u}{\partial x}+\left(\frac{\partial u}{\partial y}+\frac{\partial v}{\partial x}\right)\frac{\partial u}{\partial y}+\left(\frac{\partial v}{\partial x}+\frac{\partial u}{\partial y}\right)\frac{\partial v}{\partial x}+\left(\frac{2\partial v}{\partial y}\right)\frac{\partial v}{\partial y}\right]$$

Species continuity equation i

$$u\frac{\partial c_i}{\partial x}+v\frac{\partial c_i}{\partial y}=-\ D_i\left(\frac{\partial^2 T}{\partial x^2}+\frac{\partial^2 T}{\partial y^2}\right)+Q_i \quad [A5.5]$$

A5.1.2. *Boundary layer simplifications*

The boundary layer is "thin"; therefore, the flow is "almost parallel to the flow[1]". In Cartesian coordinates and in plane flow, we thus have a first relation between the longitudinal u and lateral v components of the velocity: $u << v$.

The longitudinal and lateral derivatives of the same function f will also have the same ratio of orders of magnitude:

$$\frac{\partial f}{\partial x} << \frac{\partial f}{\partial y} \quad [A5.6]$$

The equations of fluid mechanics will be simplified, but within limited boundaries.

Indeed, in orders of magnitude:

$$\frac{\partial u}{\partial x}\approx\frac{u}{L};\quad \frac{\partial v}{\partial y}\approx\frac{v}{\delta};\quad v<< u;\quad \delta<< L \quad [A5.7]$$

1 We will see that this cannot be done rigorously to ensure continuity. This will be the case particularly in the treatment of integral methods. In some cases, we will use this "property" in the limit in order to obtain approximate results: see, for example, the first Stokes problem that will be proposed later.

But: $\frac{u}{L} \approx \frac{v}{\delta}; \quad \frac{\partial u}{\partial x} \approx \frac{\partial v}{\partial y}$

$$u\frac{\partial u}{\partial x} \approx u\frac{u}{L}; \quad v\frac{\partial u}{\partial y} \approx u\frac{v}{\delta}; \quad \frac{\partial^2 u}{\partial y^2} \approx \frac{u}{\delta^2} \qquad [A5.8]$$

These three terms are of the same order of magnitude.

Thus, we will have the following for an incompressible and stationary plane flow over a flat plate:

$$\frac{\partial u}{\partial x} + \frac{\partial v}{\partial y} = 0 \qquad [A5.9]$$

$$u\frac{\partial u}{\partial x} + v\frac{\partial u}{\partial y} = \nu\frac{\partial^2 u}{\partial y^2} \qquad [A5.10]$$

$$u\frac{\partial T}{\partial x} + v\frac{\partial T}{\partial y} = \nu\frac{\partial^2 T}{\partial y^2} \qquad [A5.11]$$

$$u\frac{\partial c_i}{\partial x} + v\frac{\partial c_i}{\partial y} = D\frac{\partial^2 c_i}{\partial y^2} \qquad [A5.12a]$$

Note that not all derivation terms in x disappear, because these equations contain products of small terms $\left(\frac{\partial u}{\partial x}, \frac{\partial T}{\partial x}, v\right)$ by large terms $\left(u, \frac{\partial u}{\partial y}, \frac{\partial^2 u}{\partial y^2}, \frac{\partial T}{\partial y}, \frac{\partial^2 T}{\partial y^2}\right)$.

It is important to approach these notions of order of magnitude as a physicist.

In the external boundary layers, as we move away from the wall, we quickly reach a speed close to that of the potential flow, U_e. Strictly speaking, the variation is asymptotic, so we will say that u tends to U_e, whereas x tends to infinity. In wind tunnels, we will thus have infinities that are only a few cm!

Boundary layer thickness δ does not mean infinitely small.

In a wind tunnel, for a plate a few dozen centimeters long, the scale of δ is the centimeter.

If we take a ship of length $L = 250\,m$, the scale of δ will be the meter.

At the scale of a territory, the atmospheric boundary layer can measure anywhere between one and two kilometers!

Moreover, the infinitesimally small δ before L is often reduced to ratios $\frac{\delta}{L}$ in the range of 10^{-2}.

A5.1.3. *The flat plate in the wind bed: analytical solutions*

Although laminar flows may seem rather academic from a practical perspective, the reasoning developed on this model is very stimulating and instructive for the reasoning of a fluid mechanic.

We consider a fluid with constant physical properties (density μ and kinematic viscosity ν). The potential flow is then obviously a uniform flow with velocity U_e. Let us use the case of a stationary plane flow.

The system is reduced to an orthonormal reference frame. The origin is at the leading edge of the plate, the x-axis is parallel and the y-axis is normal to this plate.

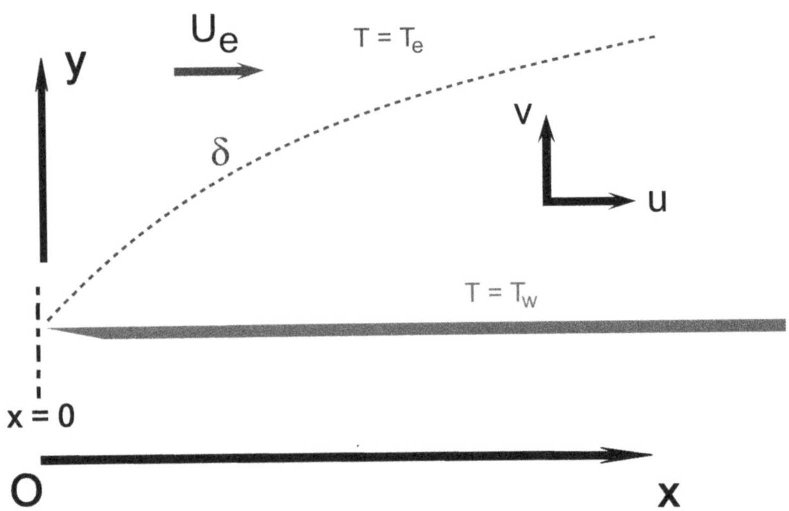

Figure A5.1. *The flat plate in the path of the wind. For a color version of this figure, see www.iste.co.uk/ledoux/heat3.zip*

A5.1.3.1. *Establishment of the Blasius and Polhausen equations*

The velocity vector $\vec{V}(u, v)$ then answers the following equations:

Continuity equation

$$\frac{\partial u}{\partial x}+\frac{\partial v}{\partial y}=0 \qquad [A5.12b]$$

Momentum equation

$$u\frac{\partial u}{\partial x}+v\frac{\partial v}{\partial y}=-\frac{1}{\rho}\frac{d\,p_G}{d\,x}+\nu\frac{\partial^2 u}{\partial y^2} \qquad [A5.13]$$

The term $\frac{d\,p_G}{d\,x}$ calls for two comments:

The flow is not strictly parallel in the boundary layer. Indeed, $v(x, y)$ is low in front of $u(x, y)$, but is not strictly zero. Physically, this is understandable: the plate "slows down" the fluid, which is "driven away from the plate" towards the free flow. However, it is close enough to a parallel flow that we can consider that, normally to these current lines, and thus to the plate, which is the lateral gradient of p_G, $\frac{\partial\,p_G}{\partial\,y}$ is zero. A *d "straight"* in x is justified there.

Moreover, in the same way, the value of $p_G(x)$ will be identical to that observed at the same abscissa x in the free flow.

Physically, U_e can vary, such as in a wind tunnel with a variable cross-section.

In the potential flow, where the fluid is considered perfect, Bernoulli's theorem holds true. Therefore:

$$\frac{d}{d\,x}\left(\rho\frac{U_e^2}{2}+p_G\right)=0 \qquad [A5.14]$$

This allows $-\frac{1}{\rho}\frac{d\,p_G}{d\,x}$ to be replaced by:

$$-\frac{1}{\rho}\frac{d\,p_G}{d\,x} = U_e\frac{d\,U_e}{d\,x} \qquad \text{[A5.15]}$$

In the case of a rigorously uniform potential flow, this term is zero, which will be the case considered in the following:

Energy equation

$$u\frac{\partial T}{\partial x} + v\frac{\partial T}{\partial y} = \nu\frac{\partial^2 T}{\partial y^2} \qquad \text{[A5.16]}$$

Blasius and Polhausen problems

Blasius solved the problem of the laminar boundary layer of a fluid with constant physical properties, over a flat plate of indefinite width, placed in the path of a wind formed by a uniform flow. Pohlhausen added a transformation of the energy equation.

The system of the three equations (continuity, momentum and energy) with three unknowns $u(x, y)$, $v(x,y)$ and $T(x,y)$ becomes:

$$\frac{\partial u}{\partial x} + \frac{\partial v}{\partial y} = 0 \qquad \text{[A5.17]}$$

$$u\frac{\partial u}{\partial x} + v\frac{\partial v}{\partial y} = -\frac{1}{\rho}\frac{d\,p_e}{d\,x} + \nu\frac{\partial^2 u}{\partial y^2} \qquad \text{[A5.18]}$$

$$u\frac{\partial T}{\partial x} + v\frac{\partial T}{\partial y} = a\frac{\partial^2 T}{\partial y^2} \qquad \text{[A5.19]}$$

The term $-\frac{1}{\rho}\frac{d\,p_e}{d\,x}$ calls for special attention. It comes from the term $-\vec{g}rad\ p$ in the general equations. A boundary layer is a flow that is very close to a parallel flow. Perpendicular to the wall, we can therefore assume a zero gradient of generated pressure $p_G = p + \rho g Z$, where the repetition of a point on a vertical axis is directed from bottom to top.

We therefore have:

$$\frac{\partial\left(p+\rho g Z\right)}{\partial y}=0 \qquad \text{[A5.20]}$$

For a non-weighing fluid (gas), the term at the front is negligible and we get the following:

The pressure is constant for a given x. It will therefore be equal to the pressure in the external flow.

Moreover, in this external flow, the fluid is assumed to be perfect. Bernoulli's theorem therefore applies. If the speed of the external flow $U_e\left(x\right)$ is variable according to x, we can write:

$$d\left(\frac{\rho U_e^2}{2}+p_e\right)=0 \qquad \text{[A5.21]}$$

This allows us to rewrite the pressure term as:

$$-\frac{1}{\rho}\frac{d\,p_e}{d\,x}=U_e\frac{d\,U_e}{d\,x} \qquad \text{[A5.22]}$$

Thus, we often find the momentum equation in the following form:

$$u\frac{\partial u}{\partial x}+v\frac{\partial v}{\partial y}\;=\;U_e\frac{d\,U_e}{d\,x}+\nu\frac{\partial^2 u}{\partial y^2} \qquad \text{[A5.23]}$$

If the external flow is uniform, then the pressure term cancels and the momentum equation becomes:

$$u\frac{\partial u}{\partial x}+v\frac{\partial v}{\partial y}\;=\;\nu\frac{\partial^2 u}{\partial y^2} \qquad \text{[A5.24]}$$

The boundary conditions being written as:

$$y=0;\quad u=v=0 \qquad \text{[A5.25]}$$

$$y\rightarrow\infty;\quad u\rightarrow U_e \qquad \text{[A5.26]}$$

We already see the symmetrical roles of ν and a from this point, now highlighting their ratio, which is the Prandtl number:

$$\mathrm{Pr} = \frac{\nu}{a} \qquad [A5.27]$$

We will note that the problem places an asymptotic variation of the velocity towards the external flow "to infinity". In fact, we will see that the convergence is fast. We define a boundary layer thickness by the point, where $u(x,\delta)$=0.99U_e. This implies that $\delta = \delta(x)$. We will verify in the following that $\frac{\delta(x)}{x}$ is low.

We will deal with the rest in the context of this last assumption.

Blasius showed that the problem becomes self-similar through the definition of a composite dimensionless variable $\eta(x,y)$:

$$\eta = y\sqrt{\frac{U_e}{\nu x}} \qquad [A5.28]$$

We perform a change of coordinates from the (x, y) system to the (ξ, η) system, with the following definitions:

$$\xi = x \qquad [A5.29]$$

$$\eta = y\sqrt{\frac{U_e}{\nu x}} \qquad [A5.30]$$

A function $f(\xi,\eta)$ then appears (which will turn out to be a unique function of η):

$$f(\xi,\eta) = \frac{\psi}{\sqrt{\nu U_e x}} \qquad [A5.31]$$

A reduced temperature $\theta(x,y)$ will also appear, defined by:

$$\theta(x,y) = \frac{T(x,y) - T_w}{T_e - T_w} \qquad [A5.32]$$

where T_e and T_w derivatives of this function $\theta(x,y)$ will be part of the form, for example, with respect to x:

$$\frac{\partial\theta(x,y)}{\partial x}=\frac{1}{T_e-T_w}\frac{\partial\theta(x,y)}{\partial x} \qquad [A5.33]$$

To do this, we introduce the current function $\Psi(x, y)$, which also allows us to remove an equation, since its definition results from the continuity equation.

By definition of the current function, the velocity components are derived by two expressions:

$$u(x,y)=\frac{\partial\psi(x,y)}{\partial y} \qquad [A5.34]$$

$$v(x,y)=-\frac{\partial\psi(x,y)}{\partial x} \qquad [A5.35]$$

The velocity then verifies the continuity equation:

$$\frac{\partial u}{\partial x}+\frac{\partial v}{\partial y}=\frac{\partial^2\psi}{\partial y\partial x}-\frac{\partial^2\psi}{\partial x\partial y}=0 \qquad [A5.36]$$

From the definition of $f(\xi,\eta)$, it becomes:

$$\psi=\sqrt{\nu U_e x}\,f(\xi,\eta)=\sqrt{\nu U_e\xi}\,f(\xi,\eta) \qquad [A5.37]$$

We will now apply this transformation to the dynamics equations, as well as to the energy equation:

$$u\frac{\partial}{\partial x}\frac{\partial\sqrt{\nu U_e x}\,f}{y}-\frac{\partial\sqrt{\nu U_e x}\,f}{\partial x}\frac{\partial}{\partial y}\frac{\partial\sqrt{\nu U_e x}\,f}{\partial y}=\nu\frac{\partial^2}{\partial y^2}\frac{\partial\sqrt{\nu U_e x}\,f}{y} \qquad [A5.38]$$

And, assuming that T_e and T_w are constant:

$$\frac{\partial\sqrt{\nu U_e x}\,f}{\partial y}\frac{\partial\theta}{\partial x}-\frac{\partial\sqrt{\nu U_e x}\,f}{\partial x}\frac{\partial\theta}{\partial y}=a\frac{\partial^2(\theta)}{\partial y^2} \qquad [A5.39]$$

Let us make the change of coordinates according to the rules:

$$\frac{\partial}{\partial x}=\frac{\partial}{\partial \xi}\frac{\partial \xi}{\partial x}+\frac{\partial}{\partial \eta}\frac{\partial \eta}{\partial x} \qquad [A5.40]$$

$$\frac{\partial}{\partial y}=\frac{\partial}{\partial \xi}\frac{\partial \xi}{\partial y}+\frac{\partial}{\partial \eta}\frac{\partial \eta}{\partial y} \qquad [A5.41]$$

with:

$$\frac{\partial \xi}{\partial x}=1 \qquad [A5.42]$$

$$\frac{\partial \xi}{\partial y}=0 \qquad [A5.43]$$

$$\frac{\partial \eta}{\partial x}=-\frac{y}{2}\sqrt{\frac{U_e}{\nu}}\,x^{-\frac{3}{2}}=-\frac{\eta}{2x} \qquad [A5.44]$$

$$\frac{\partial \eta}{\partial y}=\sqrt{\frac{U_e}{\nu x}} \qquad [A5.45]$$

which gives:

$$\frac{\partial}{\partial x}=\frac{\partial}{\partial \xi}-\frac{\eta}{2x}\frac{\partial}{\partial \eta}=\frac{\partial}{\partial \xi}-\frac{\eta}{2\xi}\frac{\partial}{\partial \eta} \qquad [A5.46]$$

$$\frac{\partial}{\partial y}=0+\sqrt{\frac{U_e}{\nu x}}\frac{\partial}{\partial \eta} \qquad [A5.47]$$

This result is:

For the momentum equation:

$$\left[\sqrt{\frac{U_e}{\nu x}}\left(\frac{\partial\sqrt{\nu U_e x}\, f}{\partial \eta}\right)\right]$$

$$\left[\frac{\partial}{\partial \xi}\sqrt{\frac{U_e}{\nu x}}\left(\frac{\partial\sqrt{\nu U_e x}\, f}{\partial \eta}\right)-\frac{\eta}{2x}\frac{\partial}{\partial \eta}\left(\sqrt{\frac{U_e}{\nu x}}\frac{\partial\sqrt{\nu U_e x}\, f}{\partial \eta}\right)\right] \qquad [A5.48]$$

$$-\left[\frac{\partial\sqrt{\nu U_e x}\, f}{\partial \xi}-\frac{\eta}{2x}\frac{\partial\sqrt{\nu U_e x}\, f}{\partial \eta}\right]\left[\sqrt{\frac{U_e}{\nu x}}\frac{\partial}{\partial \eta}\left(\sqrt{\frac{U_e}{\nu x}}\frac{\partial\sqrt{\nu U_e x}\, f}{\partial \eta}\right)\right]$$

$$=\nu\sqrt{\frac{U_e}{\nu x}}\frac{\partial}{\partial \eta}\sqrt{\frac{U_e}{\nu x}}\frac{\partial}{\partial \eta}\sqrt{\frac{U_e}{\nu x}}\frac{\partial\sqrt{\nu U_e x}\, f}{\partial \eta}$$

For the energy equation:

$$\left[\sqrt{\frac{U_e}{\nu x}}\left(\frac{\partial\sqrt{\nu U_e x}\, f}{\partial \eta}\right)\right]\left[\frac{\partial \theta}{\partial \xi}-\frac{\eta}{2x}\frac{\partial \theta}{\partial \eta}\right]$$

$$-\left[\frac{\partial\sqrt{\nu U_e x}\, f}{\partial \xi}-\frac{\eta}{2x}\frac{\partial\sqrt{\nu U_e x}\, f}{\partial \eta}\right]\left[\sqrt{\frac{U_e}{\nu x}}\frac{\partial \theta}{\partial \eta}\right]=a\sqrt{\frac{U_e}{\nu x}}\frac{\partial}{\partial \eta}\sqrt{\frac{U_e}{\nu x}}\frac{\partial \theta}{\partial \eta} \qquad [A5.49]$$

We note the following, taking $\xi = x$ into account:

$$\frac{\partial\sqrt{\nu U_e x}\, f}{\partial \xi}=\frac{\partial\sqrt{\nu U_e x}\, f}{\partial x}=\frac{1}{2}\sqrt{\frac{\nu U_e}{x}}f+\sqrt{\nu U_e x}\frac{\partial f}{\partial \xi} \qquad [A5.50]$$

which, reintroduced in the previous equations, gives:

For the momentum equation:

$$\left[\sqrt{\frac{U_e}{\nu x}}\left(\frac{\partial\sqrt{\nu U_e x}\, f}{\partial \eta}\right)\right]\left[\frac{\partial}{\partial \xi}\sqrt{\frac{U_e}{\nu x}}\left(\frac{\partial\sqrt{\nu U_e x}\, f}{\partial \eta}\right)-\frac{\eta}{2x}\frac{\partial}{\partial \eta}\left(\sqrt{\frac{U_e}{\nu x}}\frac{\partial\sqrt{\nu U_e x}\, f}{\partial \eta}\right)\right]$$

$$-\left[\frac{1}{2}\sqrt{\frac{\nu U_e}{x}}f+\sqrt{\nu U_e x}\frac{\partial f}{\partial \xi}\quad-\frac{\eta}{2x}\frac{\partial\sqrt{\nu U_e x}\,f}{\partial \eta}\right]\left[\sqrt{\frac{U_e}{\nu x}}\frac{\partial}{\partial \eta}\left(\sqrt{\frac{U_e}{\nu x}}\frac{\partial\sqrt{\nu U_e x}\,f}{\partial \eta}\right)\right]$$

$$=\nu\sqrt{\frac{U_e}{\nu x}}\frac{\partial}{\partial \eta}\sqrt{\frac{U_e}{\nu x}}\frac{\partial}{\partial \eta}\sqrt{\frac{U_e}{\nu x}}\frac{\partial\sqrt{\nu U_e x}\,f}{\partial \eta}$$

For the energy equation:

$$\left[\sqrt{\frac{U_e}{\nu x}}\left(\frac{\partial\sqrt{\nu U_e x}\,f}{\partial \eta}\right)\right]\left[\frac{\partial \theta}{\partial \xi}-\frac{\eta}{2x}\frac{\partial \theta}{\partial \eta}\right]$$

$$-\left[\frac{1}{2}\sqrt{\frac{\nu U_e}{x}}f+\sqrt{\nu U_e x}\frac{\partial f}{\partial \xi}\quad-\frac{\eta}{2x}\frac{\partial\sqrt{\nu U_e x}\,f}{\partial \eta}\right] \qquad [A5.51]$$

$$\left[\sqrt{\frac{U_e}{\nu x}}\frac{\partial \theta}{\partial \eta}\right]=a\sqrt{\frac{U_e}{\nu x}}\frac{\partial}{\partial \eta}\sqrt{\frac{U_e}{\nu x}}\frac{\partial \theta}{\partial \eta}$$

which is simplified, by introducing the following notations:

$$f_\xi=\frac{\partial f}{\partial \xi} \qquad [A5.52]$$

$$f_\eta=\frac{\partial f}{\partial \eta} \qquad [A5.53]$$

$$f_{\eta\xi}=\frac{\partial^2 f}{\partial \eta \partial \xi} \qquad [A5.54]$$

For the momentum equation:

$$U_e f_\eta\left[U_e f_{\eta\xi}-\frac{\eta}{2\xi}U_e f_{\eta\eta}\right] \qquad [A5.55]$$

$$-\left[\frac{1}{2}\sqrt{\frac{\nu U_e}{\xi}}f+\sqrt{\nu U_e \xi}\,f_\xi\quad-\frac{\eta}{2}\sqrt{\frac{\nu U_e}{\xi}}f_\eta\right]\left[\frac{U_e}{\nu x}\sqrt{\nu U_e x}\,f_{\eta\eta}\right]=\nu\frac{U_e^2}{\nu\xi}f_{\eta\eta\eta} \qquad [A5.56]$$

Rewritten again:

$$U_e^2\left[f_\eta f_{\eta\xi}-\frac{\eta}{2\xi}f_\eta f_{\eta\eta}\right]-\left[\frac{1}{2}\frac{U_e^2}{\xi}ff_{\eta\eta}+\frac{U_e^2}{\xi}f_\xi f_{\eta\eta}-\frac{\eta}{2}\frac{U_e^2}{\xi}f_\eta f_{\eta\eta}\right]$$
$$=\nu\frac{U_e^2}{\nu\xi}f_{\eta\eta\eta} \quad [A5.57]$$

What remains after simplification:

$$f_\eta f_{\eta\xi}-\frac{1}{2\xi}ff_{\eta\eta}-\frac{1}{\xi}f_\xi f_{\eta\eta}=\frac{1}{\xi}f_{\eta\eta\eta} \quad [A5.58]$$

For the energy equation:

$$U_e f_\eta\left[\theta_\xi-\frac{\eta}{2\xi}\theta_\eta\right]-\left[\frac{U_e}{2\xi}f\theta_\eta+U_e f_\xi\theta_\eta-U_e\frac{\eta}{2\xi}f_\eta\,\theta_\eta\right]=U_e\frac{a}{\nu\xi}\theta_{\eta\eta} \quad [A5.59]$$

which will simplify into:

$$f_\eta\theta_\xi-\frac{1}{2\xi}f\theta_\eta-f_\xi\theta_\eta=\frac{a}{\nu\xi}\theta_{\eta\eta} \quad [A5.60]$$

We are looking for self-similar solutions, in other words, functions f and θ that only depend on η.

We see that solutions such as $f_\xi=0$ and $\theta_\xi=0$ can exist. Let us therefore postulate self-similarity.

In practice, classic notations are used, where "∂" becomes "d straight *d*":

$$f'=\frac{df}{d\eta}=f_\eta \quad [A5.61]$$

$$f''=\frac{d^2f}{d\eta^2}=f_{\eta\eta} \quad [A5.62]$$

$$f'''=\frac{d^3f}{d\eta^3}=f_{\eta\eta\eta} \quad [A5.63]$$

$$\theta' = \frac{d\theta}{d\eta} = \theta_\eta \quad \text{[A5.64]}$$

$$\theta'' = \frac{d^2\theta}{d\eta^2} = \theta_{\eta\eta} \quad \text{[A5.65]}$$

Given $f_\xi = 0$ and $\theta_\xi = 0$, the equations become:

$$0 - \frac{1}{2\xi} ff'' - \frac{1}{\xi} 0 = \frac{1}{\xi} f''' \quad \text{[A5.66]}$$

And noting that $\Pr = \frac{\nu}{a}$ is the Prandtl number of the fluid:

$$0 - \frac{1}{2\xi} f\theta_\eta - 0 = \frac{1}{\Pr \xi}\theta_{\eta\eta} \quad \text{[A5.67]}$$

It should also be noted that:

$$f' = \frac{df}{d\eta} = \frac{d}{d\eta}\frac{\psi}{\sqrt{\nu U_e x}} = \sqrt{\frac{\nu x}{U_e}}\frac{\partial}{\partial y}\frac{\psi}{\sqrt{\nu U_e x}} = \frac{u}{U_e} \quad \text{[A5.68]}$$

Therefore, at the wall (no-slip condition):

$$f'(0) \to \frac{u(x,0)}{U_e} = 0 \quad \text{[A5.69]}$$

And away from the wall ($u \to U_e$):

$$f'(\infty) \to \frac{U_e}{U_e} = 1 \quad \text{[A5.70]}$$

The result is a new system of equations for self-similar solutions.

The Blasius equation:

$$2f''' + ff'' = 0 \quad \text{[A5.71]}$$

Assorted boundary conditions:

$$f(0) = 0; \; f'(0) = 0; \; f'(\infty) \to 1 \qquad [A5.72]$$

And for the energy equation, the Polhausen equation:

$$2\theta'' + \Pr f\theta' = 0 \qquad [A5.73]$$

Assorted boundary conditions:

$$\theta(0) = 0; \quad \theta(\infty) \to 1 \qquad [A5.74]$$

Note the symmetrical roles of f' and θ here.

A5.1.3.2. *Dynamic problem: solution to the Blasius equation*

$f'(\eta)$ is obtained in the form of a series that is calculated numerically.

There are two notable values:

$$f'(4{,}92) = 0.99 \qquad [A5.75]$$

This gives the conventional thickness $\delta(x)$ of the boundary layer:

$$\frac{u}{U_e} = 0.99 \text{ for } y = \delta \qquad [A5.76]$$

Therefore, $\eta\left[x, \delta(x)\right] = \frac{\delta}{x}\sqrt{\frac{U_e x}{\nu}} = 4.92$ [A5.77]

which we retain in the form:

$$\frac{\delta(x)}{x} = \frac{4.92}{\sqrt{\frac{U_e x}{\nu}}} = 4.92\, R_x^{0.5} \qquad [A5.78]$$

We find a boundary layer thickness increasing as $x^{0.5}$.

NOTE.– in the early publications on the subject, purists defined the boundary layer thickness as $1^{0/00}$ or $\frac{u}{U_e}=0.999$. Noting that $f'(4.95)=0.999$, the conventional boundary layer thickness became:

$$\frac{\delta(x)}{x}=\frac{4.95}{\sqrt{\frac{U_e x}{\nu}}}=4.95\,R_x^{0.5} \qquad [A5.79]$$

This is a formula that we can find. We can see that the asymptotic character of the evolution of the velocity leads to a sufficiently slow convergence, so that the definition "to 1% " is sufficient in practice.

$$f''(0)=0.33206\approx 0.332 \qquad [A5.80]$$

This allows us to calculate the friction coefficient:

$$C_f(x)=\frac{2\tau_W}{\rho U_e^2}=\frac{2\mu\left.\frac{\partial u}{\partial y}\right|_{y=0}}{\rho U_e^2}=\frac{2\mu\sqrt{\frac{U_e}{\nu x}}\frac{\partial}{\partial\eta}\left[U_e f'(0)\right]}{\rho U_e^2} \qquad [A5.81]$$

Therefore: $C_f(x)=\frac{2f''(0)}{\sqrt{\frac{U_e x}{\nu}}}=2f''(0)\,R_x^{-0.5}$ [A5.82]

where $R_x=\frac{U_e x}{\nu}$ is the axial Reynolds number.

This is a formula that we retain in the following form:

$$C_f(x)=\frac{0.664}{\sqrt{R_x}} \qquad [A5.83]$$

A5.1.3.3. *Thermal problem: solution to the Polhausen equation*

Let us use the energy equation again:

$$u\frac{\partial T}{\partial x}+v\frac{\partial T}{\partial y}=a\frac{\partial^2 T}{\partial y^2} \qquad [A5.84]$$

We have introduced a dimensionless temperature:

$$\theta = \frac{T - T_W}{T_e - T_W} \quad [A5.85]$$

We have established above that the energy equation turns into:

$$2\theta'' + \Pr f\theta' = 0 \quad [A5.86]$$

Assorted boundary conditions:

$$\theta(0) = 0; \quad \theta(\infty) \to 1 \quad [A5.87]$$

The Pohlhausen equation solution is obtained by a relatively simple integration:

$$\frac{\theta''}{\theta'} = -\frac{\Pr}{2} f \quad [A5.88]$$

Note that the Blasius equation can be rewritten as:

$$f = -\frac{2f'''}{f''} \quad [A5.89]$$

$$\frac{\theta''}{\theta'} = -\frac{\Pr}{2}\left(-\frac{2f'''}{f''}\right) \quad [A5.90]$$

$$\frac{\theta''}{\theta'} = \Pr\frac{f'''}{f''} \quad [A5.91]$$

$$Ln\,\theta' = \Pr Ln\, f'' + Ln\,C_1 \quad [A5.92]$$

$$\theta' = C_1 f''^{\Pr} \quad [A5.93]$$

$$\theta = C_1 \int_0^{\eta} \left[f''(\bar{\eta}) \right]^{\Pr} d\bar{\eta} + C_2 \quad [A5.94]$$

The constants will be determined by the boundary conditions:

$$\theta(0) = 0; \quad C_2 = 0 \quad [A5.95]$$

$$\theta(\infty)\to 1;\ C_1=\frac{1}{\int_0^\infty\left[f''(\bar{\eta})\right]^{\Pr}d\bar{\eta}} \tag{A5.96}$$

Ultimately:

$$\theta(\eta,\Pr)=\frac{\int_0^\eta\left[f''(\bar{\eta})\right]^{\Pr}d\bar{\eta}}{\int_0^\infty\left[f''(\bar{\eta})\right]^{\Pr}d\bar{\eta}} \tag{A5.97}$$

$\theta(\eta,\Pr)$ is then calculated numerically.

Going back to the definition of $\theta(x,y)=\frac{T(x,y)-T_w}{T_e-T_w}$, we can express the parietal flow density:

$$\varphi_W=-\lambda\left.\frac{\partial T}{\partial y}\right|_{y=0}-\lambda\left.\frac{\partial T}{\partial y}\right|_{y=0} \tag{A5.98}$$

Knowing that: $\frac{\partial}{\partial y}=\sqrt{\frac{U_e}{\nu x}}\frac{\partial}{\partial\eta}$ [A5.99]

$$\varphi_W=-\lambda\left.\frac{\partial T}{\partial y}\right|_{y=0}=-\lambda\left.\frac{\partial(T_e-T_w)\theta}{\partial y}\right|_{y=0}=-\lambda\sqrt{\frac{U_e}{\nu x}}(T_e-T_w)\left.\frac{\partial\theta}{\partial\eta}\right|_{\eta=0} \tag{A5.100}$$

Therefore, we must assess:

$$\left.\frac{\partial\theta}{\partial\eta}\right|_{\eta=0}=\left.\frac{\partial}{\partial\eta}\frac{\int_0^\eta\left[f''(\bar{\eta})\right]^{\Pr}d\bar{\eta}}{\int_0^\infty\left[f''(\bar{\eta})\right]^{\Pr}d\bar{\eta}}\right|_{\eta=0}=\frac{f''^{\Pr}(\bar{0})}{\int_0^\infty\left[f''(\bar{\eta})\right]^{\Pr}d\bar{\eta}}=\frac{0.332^{\Pr}(\bar{0})}{\int_0^\infty\left[f''(\bar{\eta})\right]^{\Pr}d\bar{\eta}} \tag{A5.101}$$

This term can be calculated numerically for different values of Pr.

As can be seen from the following table, the results are very well approximated by:

$$\frac{0.332^{\mathrm{Pr}}\left(\bar{0}\right)}{\int_0^{\infty}\left[f''(\bar{\eta})\right]^{\mathrm{Pr}} d\bar{\eta}} \approx 0.332\,\mathrm{Pr}^{1/3} \qquad \text{[A5.102]}$$

It is this last expression that the literature will retain.

Pr	**0.6**	**0.7**	**0.8**	**0.9**	**1**	**1.1**	**7**	**10**	**15**
$\frac{0.332^{\mathrm{Pr}}\left(\bar{0}\right)}{\int_0^{\infty}\left[f''(\bar{\eta})\right]^{\mathrm{Pr}} d\bar{\eta}}$	0.276	0.293	0.307	0.32	0.332	0.334	0.645	0.73	0.835
$0.332\ \mathrm{Pr}^{1/3}$	0.280	0.294	0.308	0.320	0.332	0.342	0.635	0.715	0.818

The expression of the flow, the convection coefficient and the Nusselt are immediately deduced:

$$\varphi_W = \lambda\sqrt{\frac{U_e}{\nu x}}\left(T_e - T_w\right)0.332\,\mathrm{Pr}^{1/3} \qquad \text{[A5.103]}$$

$$h = 0.332\,\mathrm{Pr}^{1/3}\,\lambda\sqrt{\frac{U_e}{\nu x}} \qquad \text{[A5.104]}$$

$$Nu_x = \frac{h\,x}{\lambda} = 0.332\,\mathrm{Pr}^{1/3}\,x\sqrt{\frac{U_e}{\nu x}} \qquad \text{[A5.105]}$$

Therefore, finally:

$$Nu_x = 0.332\,R_x^{0.5}\,\mathrm{Pr}^{1/3} \qquad \text{[A5.106]}$$

The Stanton number can be deduced from this:

$$St_x = \frac{h}{\rho c_P U_e} = \frac{0.332\,\mathrm{Pr}^{1/3}}{U_e} \frac{\lambda}{\rho c_P} \sqrt{\frac{U_e}{\nu x}} \qquad [A5.107]$$

Noting that:

$$\frac{\lambda}{\rho c_P} = a = \frac{a\nu}{\nu} = \nu\,\mathrm{Pr}^{-1} \qquad [A5.108]$$

$$St_x = \frac{0.332\,\mathrm{Pr}^{1/3}}{U_e} \nu\,\mathrm{Pr}^{-1} \sqrt{\frac{U_e}{\nu x}} = 0.332\,\mathrm{Pr}^{1/3}\,\mathrm{Pr}^{-1} \sqrt{\frac{\nu}{U_e x}} \qquad [A5.109]$$

Ultimately:

$$St_x = \quad 0.332\,\mathrm{Pr}^{1/3}\,R_x^{-0.5}\,\mathrm{Pr}^{-2/3} \qquad [A5.110]$$

This verifies the general relation established between the Nusselt and the Stanton.

Appendix 6

Table of Functions: $erf(x)$. $erfc(x)$ and $ierfc(x)$

The three functions *erf*(*x*), *erfc*(*x*) and *ierfc*(*x*) are fundamental in the treatment of one-dimensional unsteady conduction problems. They are defined by:

$$erf(x) = \frac{2}{\pi}\int_0^x \exp(-\xi^2)\,d\xi$$

$$erfc(x) = 1 - \frac{2}{\pi}\int_0^x \exp(-\xi^2)\,d\xi$$

$$ierfc(x) = \frac{1}{\sqrt{\pi}}\exp(-x^2) - x\,erfc(x)$$

x	*erf(x)*	*erfc(x)*	*ierfc(x)*	*x*	*erf(x)*	*erfc(x)*	*ierfc(x)*
0	0	1	0.56418958	**1**	0.84270079	0.15729921	0.05025454
0.05	0.05637198	0.94362802	0.51559947	**1.1**	0.88020507	0.11979493	0.03646538
0.1	0.11246292	0.88753708	0.46982209	**1.2**	0.91031398	0.08968602	0.02604895
0.15	0.16799597	0.83200403	0.42683646	**1.3**	0.93400794	0.06599206	0.01831432
0.2	0.22270259	0.77729741	0.38660791	**1.4**	0.95228512	0.04771488	0.01267002
0.25	0.27632639	0.72367361	0.34908866	**1.5**	0.96610515	0.03389485	0.00862286
0.3	0.32862676	0.67137324	0.31421848	**1.6**	0.97634838	0.02365162	0.00577194
0.35	0.37938205	0.62061795	0.28192557	**1.7**	0.98379046	0.01620954	0.0037993

0.4	0.42839236	0.57160764	0.25212759	**1.8**	0.9890905	0.0109095	0.00245876
0.45	0.47548172	0.52451828	0.22473278	**1.9**	0.99279043	0.00720957	0.00156419
0.5	0.52049988	0.47950012	0.19964123	**2**	0.99532227	0.00467773	0.00097802
0.55	0.56332337	0.43667663	0.17674618	**2.1**	0.99702053	0.00297947	0.00060095
0.6	0.60385609	0.39614391	0.15593537	**2.2**	0.99813715	0.00186285	0.00036282
0.66	0.64937669	0.35062331	0.13354955	**2.3**	0.99885682	0.00114318	0.0002152
0.7	0.67780119	0.32219881	0.12009827	**2.4**	0.99931149	0.00068851	0.00012539
0.75	0.71115563	0.28884437	0.10483226	**2.5**	0.99959305	0.00040695	7.1762E-05
0.8	0.74210096	0.25789904	0.09117366	**2.6**	0.99976397	0.00023603	4.0336E-05
0.85	0.77066806	0.22933194	0.07900271	**2.7**	0.99986567	0.00013433	2.2264E-05
0.9	0.79690821	0.20309179	0.06820168	**2.8**	0.99992499	7.5013E-05	1.2067E-05
0.95	0.82089081	0.17910919	0.05865589	**2.9**	0.9999589	4.1098E-05	6.4216E-06
1	0.84270079	0.15729921	0.05025454	**3**	0.99997791	2.209E-05	3.355E-06

These functions are presented in Figure A6.1.

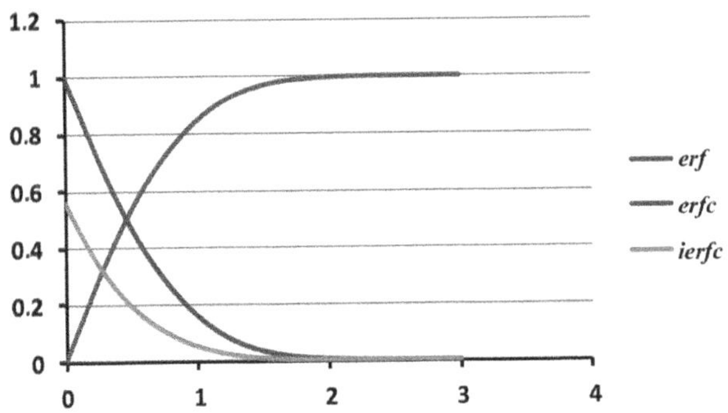

Figure A6.1. *Functions erf ξ, erfc ξ and ierfc ξ. For a color version of this figure, see www.iste.co.uk/ledoux/heat3.zip*

References

[ABO 11a] ABO AL-KHEER A., EID M., AOUES Y. et al., "Theoretical analysis of the spatial variability in tillage forces for fatigue analysis of tillage machines", *Journal of Terramechanics*, vol. 48, no. 4, pp. 285–295, 2011.

[ABO 11b] ABO AL-KHEER A., KHARMANDA M.G., EL HAMI A. et al., "Estimating the variability of tillage forces on a chisel plough shank by modeling the variability of tillage system parameters", *Computers and Electronics in Agriculture*, vol. 78, no. 1, pp. 61–70, 2011.

[ABO 11c] ABO AL-KHEER A., EL-HAMI A., KHARMANDA M.G. et al., "Reliability-based design for soil tillage machines", *Journal of Terramechanics*, vol. 48, no. 1, pp. 57–64, 2011.

[ANT 05] ANTONOVA E.E., LOOMAN D.C., "Finite elements for thermoelectric device analysis in ANSYS", *ICT 2005 24th International Conference on Thermoelectrics*, pp. 215–218, 2005.

[ASS 17] ASSI I., MJALLAL I., FARHAT H. et al., "Using phase change material in heat sinks to cool electronics devices with intermittent usage", *IEEE 7th International Conference on Power and Energy Systems (ICPES)*, pp. 66–69, 2017.

[BAB 14] BABY R., BALAJI C., "Thermal performance of a MCP heat sink under different heat loads: An experimental study", *International Journal of Thermal Sciences*, vol. 79, pp. 240–249, 2014.

[BAL 11] BALAJI C., MUNGARA P., SHARMA P., "Optimization of size and shape of composite heat sinks with phase change materials", *Heat and Mass Transfer*, vol. 47, no. 15, pp. 597–608, 2011.

[BAT 67] BATCHELOR G., *An Introduction to Fluid Dynamics*, Cambridge Mathematical Library, 1967.

[BEN 13] BEN ABDESSALEM A., PAGNACCO E., EL-HAMI A., "Increasing the stability of T-shape tube hydroforming process under stochastic framework", *International Journal of Advanced Manufacturing Technology*, vol. 69, nos 5–8, pp. 1343–1357, 2013.

[BEN 15] BEN ABDESSALEM A., EL HAMI A., "A probabilistic approach for optimising hydroformed structures using local surrogate models to control failures", *International Journal of Mechanical Sciences*, vols 96–97, pp. 143–162, 2015.

[BER 81] BERGLES A.E., COLLIER J.G., DELHAYE, J.M. et al., *Two Phase Flow and Heat Transfer in the Power and Process Industries*, Hemisphere, Washington, DC, 1981.

[BER 11] BERGMAN T.L., INCROPERA F.P., DEWITT D.P. et al., *Fundamentals of Heat and Mass Transfer*, John Wiley & Sons Inc., Hoboken, NJ, 2011.

[BIR 60] BIRD R.B., STEWART W.E., LIGHTFOOT E.N., *Transport Phenomena*, John Wiley & Sons Inc., New York, 1960.

[BIR 66] BIRD R.B., STEWART W.E., LIGHTFOOT E.N., *Transport Phenomena*, John Wiley & Sons Inc., New York, 1966.

[BOU 96] BOUYEKHF R., EL MOUDNI A., EL HAMI A. et al., "Reduced order modelling of two-time-scale discrete non-linear systems", *Journal of the Franklin Institute*, vol. 333, no. 4, pp. 499–512, 1996.

[BOU 01] BOUYEKHF R., EL HAMI A., EL MOUDNI A., "Optimal control of a particular class of singularly perturbed nonlinear discrete-time systems", *IEEE Transactions on Automatic Control*, vol. 46, no. 7, pp. 1097–1101, 2001.

[BRU 65] BRUN E.A., *Théorie de la couche limite*, Gauthier Villars, Paris, 1965.

[BRU 68] BRUN E.A., MARTINOT-LAGARDE A., MATHIEU J., *Mécanique des fluides*, Tomes I, II & III, Dunod, Paris, 1968.

[CEN 15] CENGEL Y., GHAJAR A., *Heat and Mass Transfer: Fundamentals and Applications*, 5th ed., McGraw-Hill Education, New York, pp. 1–944, 2015.

[CHE 08] CHEROUAT A., RADI B., EL HAMI A., "The frictional contact of the composite fabric's shaping", *Acta Mechanica*, vol. 199, nos 1–4, pp. 29–41, 2008.

[CHI 11] CHINTAKRINDA K., WEINSTEIN R., FLEISCHER A., "A direct comparison of three different material enhancement methods on the transient thermal response of paraffin phase change material exposed to high heat fluxes", *International Journal of Thermal Sciences*, vol. 50, no. 19, pp. 1639–1647, 2011.

[CIA 88] CIARLET P., *Introduction à l'analyse numérique matricielle et à l'optimisation*, Masson, Paris, 1988.

[COM 02] COMOLET R., *Mécanique expérimentale des fluides*, Tomes I & II, Masson, Paris, 2002.

[COS 97] COSAR P., *Aide mémoire du thermicien*, Amicale des anciens élèves de l'Ecole de thermique (ed.), Elsevier, Paris, 1997.

[COS 10] COSKUN A., ATIENZA D., ROSING T. et al., "Energy-efficient variable-flow liquid cooling in 3D stacked architectures", in *Proceedings of the Conference on Design, Automation and Test in Europe*, European Design and Automation Association, pp. 111–116, 2010.

[DAM 17] DAMMAK K., EL HAMI A., KOUBAA S. et al., "Reliability based design optimization of coupled acoustic-structure system using generalized polynomial chaos", *International Journal of Mechanical Sciences*, vol. 134, pp. 75–84, 2017.

[DAM 19] DAMMAK K., KOUBAA S., EL HAMI A. et al., "Numerical modelling of vibro-acoustic problem in presence of uncertainty: Application to a vehicle cabin", *Applied Acoustics*, vol. 144, pp. 113–123, 2019.

[DHA 84] DHATT Y., TOUZOT G., *Une présentation de la méthode des éléments finis*, Maloine, Paris, 1984.

[ECK 72] ECKERT E.R.G., DRAKE R.M., *Analysis of Heat and Mass Transfer*, McGraw-Hill, New York, 1972.

[ELH 93] EL HAMI A., LALLEMENT G., MINOTTI P. et al., "Methods that combine finite group theory with component mode synthesis in the analysis of repetitive structures", *Computers and Structures*, vol. 48, no. 6, pp. 975–982, 1993.

[ELH 96] EL HAMI A., RADI B., "Some decomposition methods in the analysis of repetitive structures", *Computers and Structures*, vol. 58, no. 5, pp. 973–980, 1996.

[ELH 09] EL HAMI A., RADI B., CHEROUAT A., "Treatment of the composite fabric's shaping using a Lagrangian formulation", *Mathematical and Computer Modelling*, vol. 49, nos 7–8, pp. 1337–1349, 2009.

[ELH 11] EL HAMI A., RADI B., "Comparison study of different reliability-based design optimization approaches", *Advanced Materials Research*, vol. 274, pp. 113–121, 2011.

[ELH 13a] EL HAMI A., RADI B., *Incertitude et optimisation et fiabilité des structures*, Hermes-Lavoisier, Paris, 2013.

[ELH 13b] EL HAMI A., RADI B., *Uncertainty and Optimization in Structural Mechanics*, ISTE Ltd, London and John Wiley & Sons, New York, 2013.

[FOK 10] FOK S., SHEN W., TAN F., "Cooling of portable hand-held electronic devices using phase change materials in finned heat sinks", *International Journal of Thermal Sciences*, vol. 49, no. 11, pp. 109–117, 2010.

[FOS 98] FOSSETT A., MAGUIRE M., KUDIRKA A. et al., "Avionics passive cooling with microencapsulated phase change materials", *Journal of Electronic Packaging*, vol. 120, no. 13, pp. 238–242, 1998.

[FRA 95] FRANC J.P., AVELLAN F., BELAHADJI B. et al., *La cavitation : mécanismes physiques et aspects industriels*, EDP Sciences, Paris, 1995.

[GOS 93] GOSSE J., *Guide technique de thermique*, Dunod, Paris, 1993.

[GUE 15a] GUERINE A., EL HAMI A., WALHA L. et al., "Approach for the dynamic analysis of one stage gear system with uncertain parameters", *Mechanism and Machine Theory*, vol. 92, pp. 113–126, 2015.

[GUE 15b] GUERINE A., EL HAMI A., FAKHFAKH T. et al., "A polynomial chaos method to the analysis of the dynamic behavior of spur gear system", *Structural Engineering and Mechanics*, vol. 53, no. 4, pp. 819–831, 2015.

[GUE 16a] GUERINE A., EL HAMI A., WALHA L. et al., "A polynomial chaos method for the analysis of the dynamic behavior of uncertain gear friction system", *European Journal of Mechanics – A/Solids*, vol. 59, pp. 76–84, 2016.

[GUE 16b] GUERINE A., HAMI A.E., WALHA L. et al., "Dynamic response of a Spur gear system with uncertain friction coefficient", *Advances in Engineering Software*, vol. 120, pp. 45–54, 2016.

[GUE 17] GUERINE A., EL HAMI A., WALHA L. et al., "Dynamic response of wind turbine gear system with uncertain-but-bounded parameters using interval analysis method", *Renewable Energy*, vol. 113, pp. 679–687, 2017.

[HAR 56] HARTNETT E., BIRKEBAK S., Simplified procedure for the calculation of heat transfer to surfaces with non-uniform temperature, WADC report 56–373, December 1956.

[HAS 16] HASAN A., HEJASE H., ABDELBAQI S. et al., "Comparative effectiveness of different phase change materials to improve cooling performance of heat sinks for electronic devices", *Applied Sciences*, vol. 6, no. 19, p. 226, 2016.

[HEL 03] HELTON J.C., DAVIS F.J., "Latin hypercube sampling and the propagation of uncertainty in analyses of complex systems", *Reliability Engineering & System Safety*, vol. 81, pp. 23–69, July 2003.

[HER 55] HERMANN S., *Boundary Layer Theory*, Springer Verlag, Berlin, Heidelberg, 1955.

[HER 17] HERMANN S., *Boundary Layer Theory*, Springer Verlag, Berlin, Heidelberg, 2017.

[HIN 75] HINZE J.O., *Turbulence*, McGraw Hill, New York, 1975.

[HOD 02] HODES M., WEINSTEIN R.D., PENCE S.J. et al., "Transient thermal management of a handset using phase change material (MCP)", *Journal of Electronic Packaging*, vol. 124, no. 14, pp. 419–426, 2002.

[HOL 86] HOLMAN J.P., *Heat Transfer*, McGraw Hill, New York, 1986.

[HUA 16] HUANG C., RADI B., EL HAMI A., "Uncertainty analysis of deep drawing using surrogate model based probabilistic method", *Manufacturing Technology*, vol. 86, nos 9–12, pp. 3229–3240, 2016.

[JAN 12] JANNOT Y., MOYNE C., Transferts thermique, Ecole des Mines de Nancy, 2012.

[JAN 17] JANNOUN M., AOUES Y., PAGNACCO E. et al., "Probabilistic fatigue damage estimation of embedded electronic solder joints under random vibration", *Microelectronics Reliability*, vol. 78, pp. 249–257, 2017.

[KAN 07] KANDASAMY R., WANG X., MUJUMDAR A., "Application of phase change materials in thermalmanagement of electronics", *Applied Thermal Engineering*, vol. 27, nos 117–118, pp. 2822–2832, 2007.

[KAN 08] KANDASAMY R., WANG X., MUJUMDAR A., "Transient cooling of electronics using phase change material", *Applied Thermal Engineering*, pp. 1047–1057, 2008.

[KRI 04] KRISHNAN S., GARIMELLA S.V., "Analysis of a phase change energy storage system for pulsed power dissipation", *IEEE Transactions on Components and Packaging Technologies*, vol. 27, no. 11, pp. 191–199, 2004.

[LAN 59] LANDAU L., LIFCHITZ E., *Fluid Mechanics*, Pergamon Press, London, 1959.

[LAN 84] LANDAU L., LIFSHITZ E.M., *Electrodynamics of Continuous Media*, 2nd ed., Butterworth-Heinemann, Oxford, 1984.

[LAN 94] LANDAU L., LIFCHITZ E., *Mécanique des Fluides*, Ellipses, Paris, Editions MIR, Moscow, 1994.

[LED 75] LEDOUX M., Simulation thermique d'écoulements avec réaction de surface. Etude de la couche limite de diffusion en régime laminaire et turbulent, PhD thesis, Université de Rouen, February 1975.

[LED 17a] LEDOUX M., EL HAMI A., *Fluid Mechanics Analytical Methods*, ISTE Ltd, London and John Wiley & Sons, New York, 2017.

[LED 17b] LEDOUX M., EL HAMI A., *Fluid Mechanics, Compressible Flow, Propulsion and Digital Approaches in Fluid Dynamics*, ISTE Ltd, London, John Wiley & Sons, New York, 2017.

[LED 21a] LEDOUX M., EL HAMI A., *Heat Transfer 1: Conduction*, ISTE Ltd, London, John Wiley & Sons, New York, 2021.

[LED 21b] LEDOUX M., EL HAMI A., *Heat Transfer 2: Radiative Transfer*, ISTE Ltd, London, John Wiley & Sons, New York, 2021.

[LEO 79] LEONTIEV A., *Théorie des échanges de chaleur et de masse*, Editions MIR, Moscow, 1979.

[LEO 00] LEONI N., AMON C., "Bayesian surrogates for integrating numerical, analytical, and experimental data: Application to inverse heat transfer in wearable computers", *IEEE Transactions on Components and Packaging Technologies*, vol. 23, pp. 23–32, 2000.

[LIG 50] LIGHTHILL M.J., "Contribution to the theory of heat transfer through a laminar boundary layer", *Proceedings of the Royal Society A*, pp. 202–359, 1950.

[MAB 17] MABROUK I.B., EL HAMI A., WALHA L. et al., "Dynamic vibrations in wind energy systems: Application to vertical axis wind turbine", *Mechanical Systems and Signal Processing*, vol. 85, pp. 396–414, 2017.

[MAK 16] MAKHLOUFI A., AOUES Y., EL HAMI A., "Reliability based design optimization of wire bonding in power microelectronic devices", *Microsystem Technologies*, vol. 22, no. 12, pp. 2737–2748, 2016.

[MJA 18] MJALLAL I., FARHAT H., HAMMOUD M. et al., "Improving the cooling efficiency of heat sinks through the use of different types of phase change materials", *Technologies*, vol. 1, no. 15, p. 6, 2018.

[MOR 02] MORO T., EL HAMI A., EL MOUDNI A., "Reliability analysis of a mechanical contact between deformable solids", *Probabilistic Engineering Mechanics*, vol. 17, no. 3, pp. 227–232, 2002.

[NIS 03] NISTOR I., PANTALÉ O., CAPERAA S. et al., "Identification of a dynamic viscoplastic flow law using a combined Levenberg-Marquardt and Monte-Carlo algorithm", *VII International Conference on Computational Plasticity COMPLAS*, Barcelona, 2003.

[NYE 57] NYE J.F., *Physical Properties of Crystals: Their Representation by Tensors and Matrices*, Clarendon Press, Oxford, 1957.

[PAL 98] PAL D., JOSHI Y., "Thermal management of an avionics module using solid-liquid phase-change materials", *Journal of Thermophysics and Heat Transfer*, vol. 12, no. 12, pp. 256–262, 1998.

[PER 97] PEREZ J.P., *Thermodynamique, fondements et applications*, Masson, Paris, 1997.

[PY 01] PY X., OLIVES R., MAURAN S., "Paraffin/porous-graphite-matrix composite as a high and constant power thermal storage material", *International Journal of Heat and Mass Transfer*, vol. 44, no. 114, pp. 2727–2737, 2001.

[RAD 07] RADI B., EL HAMI A., "Reliability analysis of the metal forming process", *Mathematical and Computer Modelling*, vol. 45, nos 3–4, pp. 431–439, 2007.

[ROH 98] ROHSENOW W.M., HARTNETT J.P., CHO Y.I., *Handbook of Heat Transfer*, McGraw Hill, New York, 1998.

[RUB 51] RUBESIN M.W., The effect of arbitrary surface temperature variation along a flat plate on the convective heat transfer in an incompressible turbulent boundary layer, NACA TN 2345, 1951.

[SAC 15] SACCADURA J.F., *Transferts thermiques, initiation et approfondissement*, Lavoisier, Paris, 2015.

[SAR 11] SARSRI D., AZRAR L., JEBBOURI A. et al., "Component mode synthesis and polynomial chaos expansions for stochastic frequency functions of large linear FE models", *Computers and Structures*, vol. 89, nos 3–4, pp. 346–356, 2011.

[SEB 51] SEBAN R.A., Experimental investigation of convective heat transfer on a pipe to air from a flat plate with stepwise discontinuous surface temperature, Thesis, University of Califormnia, Berkeley, 1951.

[SHI 72] SHINOZUKA M., "Monte Carlo solution of structural dynamics", *Computers & Structures*, vol. 2, pp. 855–874, 1972.

[SUS 11] SUSMAN G., DEHOUCHE Z., CHEECHERN T. et al., "Tests of prototype MCP 'sails' for office cooling", *Applied Thermal Engineering*, vol. 31, no. 15, pp. 717–726, 2011.

[TIA 11] TIAN Y., ZHAO C.Y., "A numerical investigation of heat transfer in phase change materials (MCPs) embedded in porous metals", *Energy*, vol. 36, no. 19, pp. 5539–5546, 2011.

[TEN 19] TENNEKES H., LUMLEY J.L., *A First Course in Turbulence*, MIT Press, Cambridge, MA, 2019.

[TRI 72] TRINITÉ M., VALENTIN, P., "Turbulent boundary layer with discontinuity in wall temperature and concentration", *International Journal of Heat and Mass Transfer*, vol. 15, no. 7, pp. 1337–1354, July 1972.

[YAN 09] YANG Y.T., PENG H.S., "Investigation of planted pin fins for heat transfer enhancement in plate fin heat sink", *Microelectronics Reliability*, vol. 49, no. 12, pp. 163–169, 2009.

[ZHE 04] ZHENG N., WIRTZ R.A., "A hybrid thermal energy storage device, part 1: Design methodology", *Journal of Electronic Packaging*, vol. 126, no. 11, pp. 1–7, 2004.

[ZHO 11] ZHOU D., ZHAO C.Y., "Experimental investigations on heat transfer in phase change materials (MCPs) embedded in porous materials", *Applied Thermal Engineering*, vol. 31, no. 15, pp. 970–977, 2011.

Index

R, S

T, V

Other titles from

iSTE

in

Mechanical Engineering and Solid Mechanics

2022

BAYLE Franck
Product Maturity 1: Theoretical Principles and Industrial Applications (Reliability of Multiphysical Systems Set – Volume 12)
Product Maturity 2: Principles and Illustrations (Reliability of Multiphysical Systems Set – Volume 13)

2021

CHALLAMEL Noël, KAPLUNOV Julius, TAKEWAKI Izuru
Modern Trends in Structural and Solid Mechanics 1: Static and Stability
Modern Trends in Structural and Solid Mechanics 2: Vibrations
Modern Trends in Structural and Solid Mechanics 3: Non-deterministic Mechanics

DAHOO Pierre Richard, POUGNET Philippe, EL HAMI Abdelkhalak
Applications and Metrology at Nanometer Scale 1: Smart Materials, Electromagnetic Waves and Uncertainties (Reliability of Multiphysical Systems Set – Volume 9)

Applications and Metrology at Nanometer Scale 2: Measurement Systems, Quantum Engineering and RBDO Method (Reliability of Multiphysical Systems Set – Volume 10)

LEDOUX Michel, EL HAMI Abdelkhalak
Heat Transfer 1: Conduction (Mathematical and Mechanical Engineering Set – Volume 9)
Heat Transfer 2: Radiative transfer (Mathematical and Mechanical Engineering Set – Volume 10)

PLANCHETTE Guy
Cindynics, The Science of Danger: A Wake-up Call (Reliability of Multiphysical Systems Set – Volume 11)

2020

SALENÇON Jean
Elastoplastic Modeling

2019

BAYLE Franck
Reliability of Maintained Systems Subjected to Wear Failure Mechanisms: Theory and Applications
(Reliability of Multiphysical Systems Set – Volume 8)

BEN KAHLA Rabeb, BARKAOUI Abdelwahed, MERZOUKI Tarek
Finite Element Method and Medical Imaging Techniques in Bone Biomechanics (Mathematical and Mechanical Engineering Set – Volume 8)

IONESCU Ioan R., QUEYREAU Sylvain, PICU Catalin R., SALMAN Oguz Umut
Mechanics and Physics of Solids at Micro- and Nano-Scales

LE VAN Anh, BOUZIDI Rabah
Lagrangian Mechanics: An Advanced Analytical Approach

MICHELITSCH Thomas, PÉREZ RIASCOS Alejandro, COLLET Bernard, NOWAKOWSKI Andrzej, NICOLLEAU Franck
Fractional Dynamics on Networks and Lattices

SALENÇON Jean
Viscoelastic Modeling for Structural Analysis

VÉNIZÉLOS Georges, EL HAMI Abdelkhalak
Movement Equations 5: Dynamics of a Set of Solids
(Non-deformable Solid Mechanics Set – Volume 5)

2018

BOREL Michel, VÉNIZÉLOS Georges
Movement Equations 4: Equilibriums and Small Movements
(Non-deformable Solid Mechanics Set – Volume 4)

FROSSARD Etienne
Granular Geomaterials Dissipative Mechanics: Theory and Applications in Civil Engineering

RADI Bouchaib, EL HAMI Abdelkhalak
Advanced Numerical Methods with Matlab® 1: Function Approximation and System Resolution
(Mathematical and Mechanical Engineering SET – Volume 6)
Advanced Numerical Methods with Matlab® 2: Resolution of Nonlinear, Differential and Partial Differential Equations
(Mathematical and Mechanical Engineering SET – Volume 7)

SALENÇON Jean
Virtual Work Approach to Mechanical Modeling

2017

BOREL Michel, VÉNIZÉLOS Georges
Movement Equations 2: Mathematical and Methodological Supplements
(Non-deformable Solid Mechanics Set – Volume 2)
Movement Equations 3: Dynamics and Fundamental Principle
(Non-deformable Solid Mechanics Set – Volume 3)

BOUVET Christophe
Mechanics of Aeronautical Solids, Materials and Structures
Mechanics of Aeronautical Composite Materials

BRANCHERIE Delphine, FEISSEL Pierre, BOUVIER Salima, IBRAHIMBEGOVIC Adnan
From Microstructure Investigations to Multiscale Modeling: Bridging the Gap

CHEBEL-MORELLO Brigitte, NICOD Jean-Marc, VARNIER Christophe
From Prognostics and Health Systems Management to Predictive Maintenance 2: Knowledge, Traceability and Decision
(Reliability of Multiphysical Systems Set – Volume 7)

EL HAMI Abdelkhalak, RADI Bouchaib
Dynamics of Large Structures and Inverse Problems
(Mathematical and Mechanical Engineering Set – Volume 5)
Fluid-Structure Interactions and Uncertainties: Ansys and Fluent Tools
(Reliability of Multiphysical Systems Set – Volume 6)

KHARMANDA Ghias, EL HAMI Abdelkhalak
Biomechanics: Optimization, Uncertainties and Reliability
(Reliability of Multiphysical Systems Set – Volume 5)

LEDOUX Michel, EL HAMI Abdelkhalak
Compressible Flow Propulsion and Digital Approaches in Fluid Mechanics
(Mathematical and Mechanical Engineering Set – Volume 4)
Fluid Mechanics: Analytical Methods
(Mathematical and Mechanical Engineering Set – Volume 3)

MORI Yvon
Mechanical Vibrations: Applications to Equipment

2016

BOREL Michel, VÉNIZÉLOS Georges
Movement Equations 1: Location, Kinematics and Kinetics
(Non-deformable Solid Mechanics Set – Volume 1)

BOYARD Nicolas
Heat Transfer in Polymer Composite Materials

CARDON Alain, ITMI Mhamed
New Autonomous Systems
(Reliability of Multiphysical Systems Set – Volume 1)

DAHOO Pierre Richard, POUGNET Philippe, EL HAMI Abdelkhalak
Nanometer-scale Defect Detection Using Polarized Light
(Reliability of Multiphysical Systems Set – Volume 2)

DE SAXCÉ Géry, VALLÉE Claude
Galilean Mechanics and Thermodynamics of Continua

DORMIEUX Luc, KONDO Djimédo
Micromechanics of Fracture and Damage
(Micromechanics Set – Volume 1)

EL HAMI Abdelkhalak, RADI Bouchaib
Stochastic Dynamics of Structures
(Mathematical and Mechanical Engineering Set – Volume 2)

GOURIVEAU Rafael, MEDJAHER Kamal, ZERHOUNI Noureddine
From Prognostics and Health Systems Management to Predictive Maintenance 1: Monitoring and Prognostics
(Reliability of Multiphysical Systems Set – Volume 4)

KHARMANDA Ghias, EL HAMI Abdelkhalak
Reliability in Biomechanics
(Reliability of Multiphysical Systems Set –Volume 3)

MOLIMARD Jérôme
Experimental Mechanics of Solids and Structures

RADI Bouchaib, EL HAMI Abdelkhalak
Material Forming Processes: Simulation, Drawing, Hydroforming and Additive Manufacturing
(Mathematical and Mechanical Engineering Set – Volume 1)

2015

KARLIČIĆ Danilo, MURMU Tony, ADHIKARI Sondipon, MCCARTHY Michael
Non-local Structural Mechanics

SAB Karam, LEBÉE Arthur
Homogenization of Heterogeneous Thin and Thick Plates

2014

ATANACKOVIC M. Teodor, PILIPOVIC Stevan, STANKOVIC Bogoljub, ZORICA Dusan
Fractional Calculus with Applications in Mechanics: Vibrations and Diffusion Processes
Fractional Calculus with Applications in Mechanics: Wave Propagation, Impact and Variational Principles

CIBLAC Thierry, MOREL Jean-Claude
Sustainable Masonry: Stability and Behavior of Structures

ILANKO Sinniah, MONTERRUBIO Luis E., MOCHIDA Yusuke
The Rayleigh–Ritz Method for Structural Analysis

LALANNE Christian
Mechanical Vibration and Shock Analysis – 5-volume series – 3rd edition
Sinusoidal Vibration – Volume 1
Mechanical Shock – Volume 2
Random Vibration – Volume 3
Fatigue Damage – Volume 4
Specification Development – Volume 5

LEMAIRE Maurice
Uncertainty and Mechanics

2013

ADHIKARI Sondipon
Structural Dynamic Analysis with Generalized Damping Models: Analysis

ADHIKARI Sondipon
Structural Dynamic Analysis with Generalized Damping Models: Identification

BAILLY Patrice
Materials and Structures under Shock and Impact

BASTIEN Jérôme, BERNARDIN Frédéric, LAMARQUE Claude-Henri
Non-smooth Deterministic or Stochastic Discrete Dynamical Systems: Applications to Models with Friction or Impact

EL HAMI Abdelkhalak, RADI Bouchaib
Uncertainty and Optimization in Structural Mechanics

KIRILLOV Oleg N., PELINOVSKY Dmitry E.
Nonlinear Physical Systems: Spectral Analysis, Stability and Bifurcations

LUONGO Angelo, ZULLI Daniele
Mathematical Models of Beams and Cables

SALENÇON Jean
Yield Design

2012

DAVIM J. Paulo
Mechanical Engineering Education

DUPEUX Michel, BRACCINI Muriel
Mechanics of Solid Interfaces

ELISHAKOFF Isaac *et al.*
Carbon Nanotubes and Nanosensors: Vibration, Buckling and Ballistic Impact

GRÉDIAC Michel, HILD François
Full-Field Measurements and Identification in Solid Mechanics

GROUS Ammar
Fracture Mechanics – 3-volume series
Analysis of Reliability and Quality Control – Volume 1
Applied Reliability – Volume 2
Applied Quality Control – Volume 3

RECHO Naman
Fracture Mechanics and Crack Growth

2011

KRYSINSKI Tomasz, MALBURET François
Mechanical Instability

SOUSTELLE Michel
An Introduction to Chemical Kinetics

2010

BREITKOPF Piotr, FILOMENO COELHO Rajan
Multidisciplinary Design Optimization in Computational Mechanics

DAVIM J. Paulo
Biotribolgy

PAULTRE Patrick
Dynamics of Structures

SOUSTELLE Michel
Handbook of Heterogenous Kinetics

2009

BERLIOZ Alain, TROMPETTE Philippe
Solid Mechanics using the Finite Element Method

LEMAIRE Maurice
Structural Reliability

2007

GIRARD Alain, ROY Nicolas
Structural Dynamics in Industry

GUINEBRETIÈRE René
X-ray Diffraction by Polycrystalline Materials

KRYSINSKI Tomasz, MALBURET François
Mechanical Vibrations

KUNDU Tribikram
Advanced Ultrasonic Methods for Material and Structure Inspection

SIH George C. *et al.*
Particle and Continuum Aspects of Mesomechanics

Printed and bound by CPI Group (UK) Ltd, Croydon, CR0 4YY
16/09/2026
14963083-0005